The Complete Study Guide to Learning the Electrocardiogram

AF572462

Douglas Martin MSN RN

1st Edition

Disclaimer

The purpose of this self-learning module is to assist the practitioner in becoming more comfortable and familiar with interpreting the electrocardiogram and recognizing EKG rhythms that are detrimental or lethal to the patient. The clinical assessments and treatments described in this manuscript are based on research from nursing text, medical text and consulting with healthcare professionals. The author and publisher disclaim any responsibility for any adverse effects that may result from suggested procedures, undetected errors or from the readers misunderstanding of the text.

Copyright Notice

© 2007
Mind-Forge Publishing
Surprise, Arizona.
http://Mind-Forge.com/

The software, graphics and content of this educational packet are copyrighted and protected materials. No portion of this packet or software may be copied, redistributed or electronically transferred without the expressed written permission of Mind-Forge Education Services. Receipt of Copyright is on file with the United States Library of Congress.

Library of Congress Cataloging-in-Publication data

Martin, Douglas D.

The compete study guide to learning the electrocardiogram / Douglas Martin 1st Edition.
Includes bibliographic references and table contents.
TXu1-316-83
ISBN 978-0-6151-4214-2

Dedicated with
love to my wife, Sandra.

Instructions to the Learner

This self-learning instructional module is intended to assist the clinician in learning and understanding the electrocardiogram. To properly understand and interpret the electrocardiogram with competence requires an understanding of the patient, cardiac anatomy and physiology, practice and more practice. You should seek out every learning opportunity that is presented to you in the clinical setting. While this manuscript contains many of the common electrocardiogram rhythms that you may expect to encounter, unique patients will present with some things that you will not see in a textbook. It is these unique experiences that a textbook cannot teach you, and which you will only gain with hands on clinical experience. Seek out learning opportunities; do not wait for them to find you.

You should read and study the materials, and when ready, move onto the workbook section of this text. You should use the workbook to learn how to properly measure the different intervals and waveforms, properly calculate heart rate and correctly identify the different electrocardiogram rhythms. The answer key should be utilized to check your work. When you are ready to move beyond the textbook and into the clinical setting, there is a pocket guide in the back of this text, which you can keep handy and use it to quickly review your electrocardiogram rhythms.

Table of Contents

Objectives

After reading the text and completing the workbook, the learner should be able to:

- Identify the components of the cardiac conduction system.
- Discuss electrical anatomy and physiology of the heart.
- Identify individual waveform components of the EKG.
- Identify normal criteria for each of the EKG waveforms and interval measurement criteria.
- Describe correct placement of 3 and 5 lead EKG electrode placement to perform electrocardiography.
- Calculate the heart rate of regular and irregular EKG rhythms.
- Describe the characteristics of a normal EKG complex.
- Identify abnormal EKG rhythms.
- Identify lethal EKG rhythms.
- Correctly analyze a variety of EKG rhythms strips contained in the EKG workbook by identifying the rhythm, rate and intervals.

EKG Graph Paper

The paper used to record EKG tracings is grid like in nature and has specific markings utilized to mark length in seconds. The paper is divided into large and small boxes. Thicker darker lines separate large boxes; thinner lighter lines separate small boxes.

- The dark vertical lines are 0.20 seconds apart.
- The lighter vertical lines are 0.04 seconds apart.
- Each small box is 1mm in size.
- Each small box represents an electrical current that is equal to 0.1 (millivolt) mV.
- One mV is equal to two large boxes.

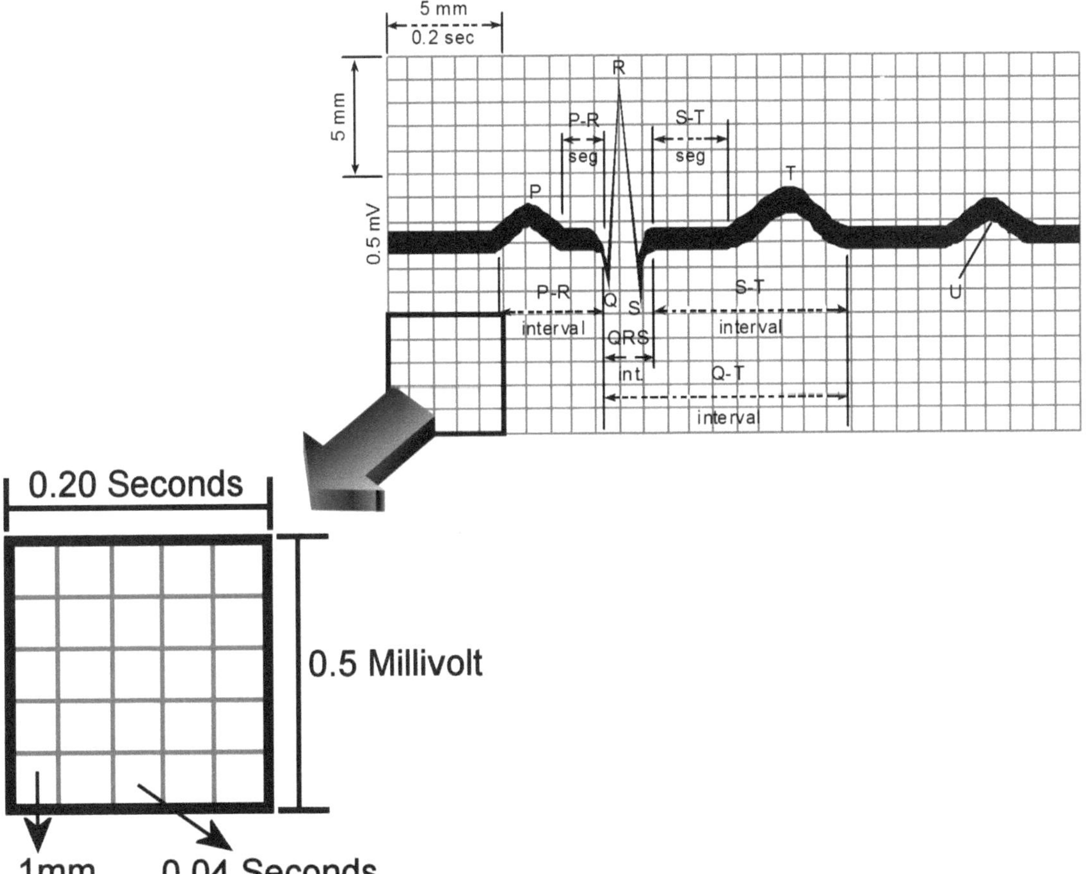

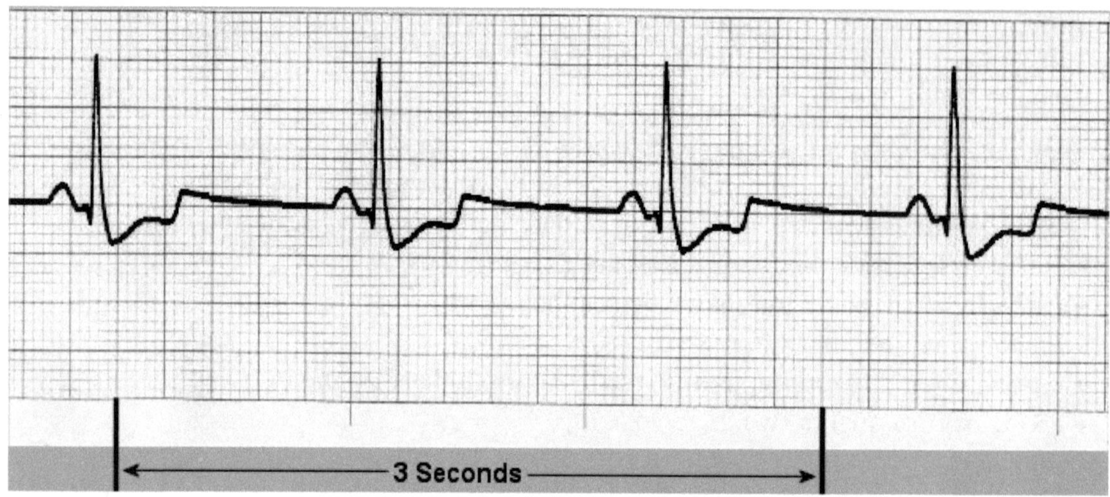

Normally EKG graph paper has thick dark lines at the top or bottom of the paper. In the strip presented here, the lines are at the bottom of the paper. The space between the lines is equal to 3 seconds. A standard EKG strip is run over 6 seconds. When printing an EKG rhythm strip the paper prints at a standard speed of 25 mm per second.

Components of the Normal EKG Waveform

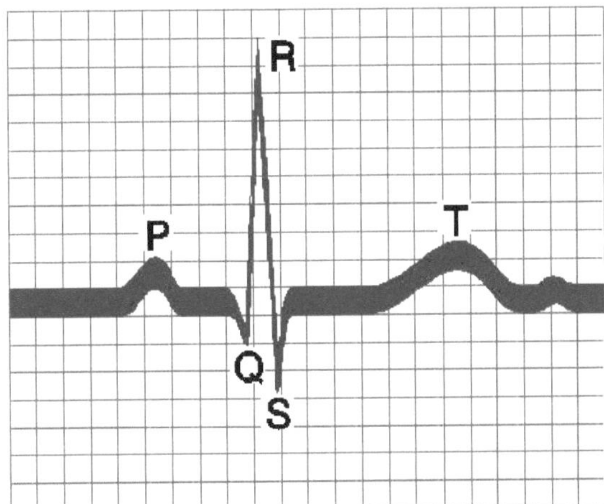

The EKG is a real-time recording of the heart's electrical activity, produced by depolarization and repolarization of the hearts cells.

The EKG waveform consists of the following waves:

- **P Wave:** Represents atrial depolarization.
- **QRS:** Represents ventricular depolarization.
- **T Wave:** Represents ventricular repolarization.

Intervals are described as the length of time between one waveform and the next. Waveforms and intervals are described in detail below.

Isoelectric Line: Flat line of the EKG tracing that represents no electrical activity, and is often referred to as the baseline. Deflections above the isoelectric line are positive, and deflections below the isoelectric line are negative.

P-Wave: The SA node produces an electrical stimulus that causes the atria to depolarize and contract in response to the stimulus. The P wave is normally upright (in lead II), symmetrical with no notching or peaking and is usually no more than 3 mm in height.

PRI (PR Interval): Is representative of the spread of the atrial depolarization wave, and the time it takes for the impulse to conduct through the AV node and to the ventricles. The PRI is marked from the start of the P wave to the beginning of the QRS. The PRI is normally no more than 0.20 seconds in length. Any more than 0.20 indicates an abnormality in the conduction system. When the SA impulse arrives at the AV node, there is a normal pause of 0.10 seconds that allows the atria to fully depolarize and contract. Most of the PRI delay is spent in the AV node, as it serves as a safety mechanism to allow for full atrial depolarization and to filter out rapid impulses and other electrical abnormalities that might prove dangerous to the rest of the heart.

QRS Complex: The QRS represents depolarization of the ventricles and ventricular conduction time of the electrical impulse. The term QRS is often used to generically describe EKG waveforms that originate from the ventricles. Typically, the QRS is narrow with a conduction time of no more than 0.12 seconds. The QRS is measured from the beginning of the first waveform to the point at which the waveform returns to the isoelectric line. A conduction time greater than 0.12 seconds may represent abnormal conduction and warrants further investigation and scrutiny on the part of the clinician. The QRS is made up of the following waves:

- **Q-Wave:** First negative deflection from baseline (below the isoelectric line).
- **R-Wave:** Positive deflection from the baseline (above the isoelectric line).
- **S-Wave:** Negative deflection following the R wave (below the isoelectric line).
- **R' (Prime)** is a secondary positive wave that may represent abnormal ventricular conduction.

Below are examples of the shapes that the QRS complex may take:

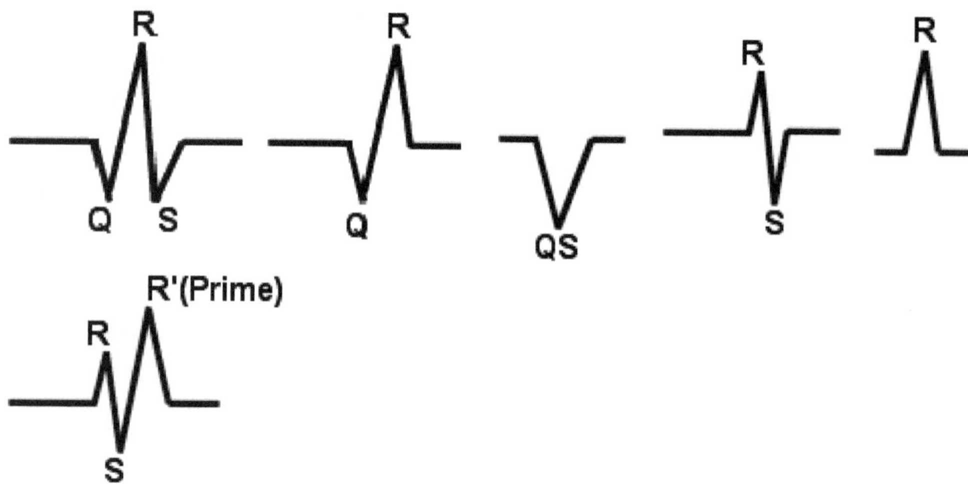

ST-Segment: Represents early ventricular repolarization, and extends from the end of the QRS to the beginning of the T wave. Normally the ST-Segment is even with the isoelectric line. A deviation either above or below the isoelectric line represents myocardial injury or ischemia. This warrants further investigation and scrutiny by the clinician.

T-Wave: Represents ventricular repolarization as the ventricles return to a state of relaxation. The T wave is typically rounded and systematical. The T wave is typically upright in lead II. However, this may vary if myocardial injury or ischemia is present.

QT-Interval (QTI): The QTI represents the refractory period of the ventricles, as they depolarize and repolarize. As rule of thumb, a normal QTI is less than 0.40 seconds. The QTI is directly related to the heart rate. As the heart beats faster, the QTI shortens as the heart repolarizes/depolarizes at a faster rate. As the heart slows down, it can depolarize/repolarize at a slower rate, this is noted by a longer, slower QTI. A prolonged QTI places the heart at risk for an R-ON-T phenomenon, this is discussed later.

Summary of Heart Measurements:

- **P-Wave:** Upright in lead II.
- **PRI:** < 0.20 Sec.
- **QRS:** < 0.12 Sec.
- **ST-Segment:** Even with isoelectric line.
- **T-Wave:** Upright in lead II.
- **QTI:** < 0.40 Sec.

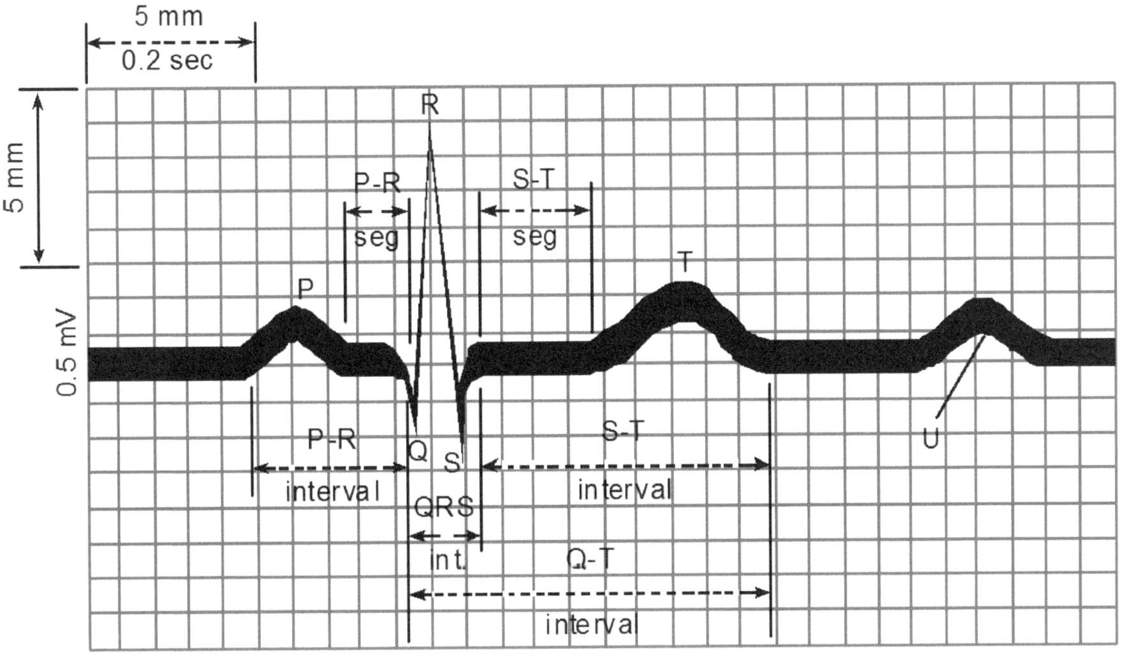

Basic Cardiac Electrophysiology

Basic Definitions Related to Electrophysiology of the Cardiac System

Automaticity: The ability of cardiac cells to spontaneously initiate an electrical impulse without external stimulation.

Conductivity: The ability of cardiac cells to propagate an electrical impulse from cell to cell.

Conducting Cells: Specialized cells of the conduction system of the heart that are capable of initiating an impulse and rapidly spreading the impulse in an organized manner throughout the heart. These cells do not contract, but are capable of rapidly conducting electrical impulses.

Contractility: The ability of cardiac muscle cells to contract (shorten and lengthen) in response to electrical stimulation.

Contracting Cells: These cells make up the walls of the atria and walls of the ventricles, and are responsible for contraction of those chambers. Contracting cells must receive an electrical stimulus before they will contract. When one cell is stimulated, the impulse is spread to all cells in the chamber, resulting in all the cells contracting as a single unit.

Depolarization: Electrical excitation of the cell membrane resulting from the flow of ions across the cell's membrane. The electrical impulse spreads through out the conduction system and muscle fibers of the heart, providing the stimulus to contract.

Excitability: The ability of cardiac cells to reach a threshold and respond to a stimulus.

Refractory Period: The period of time after a cell has depolarized during which it cannot depolarize again until it has partially or fully repolarized.

Repolarization: The return of the cell membrane to its resting potential, due to the flow of ions across the cell membrane. A cell must repolarize before it can respond to a stimulus.

Cardiac Conduction System

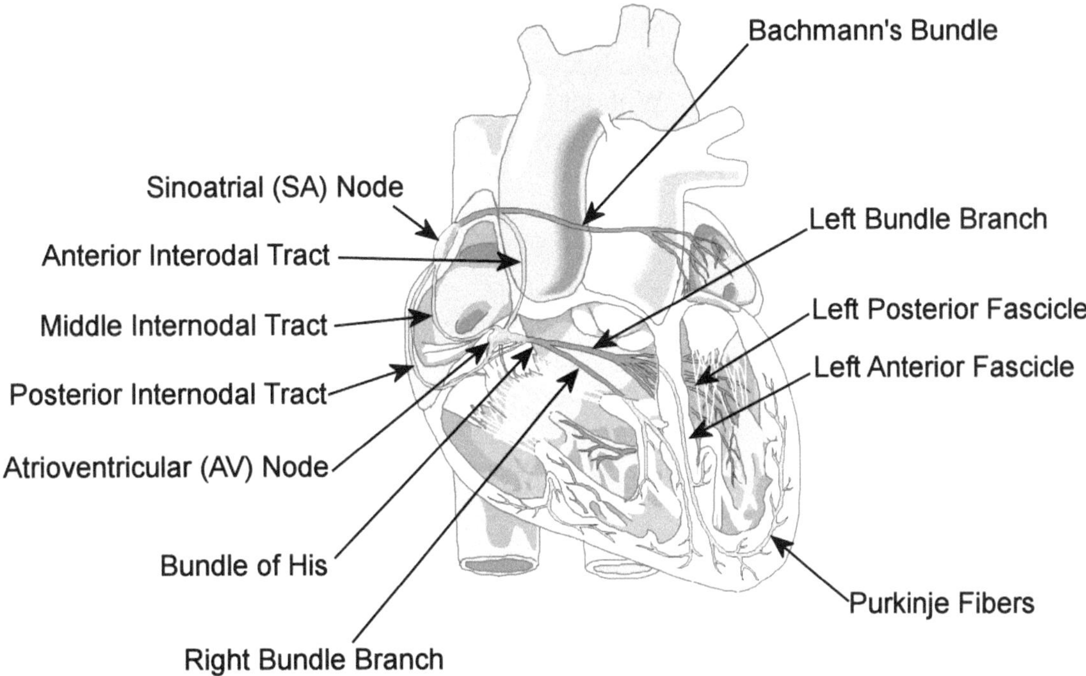

Sinus Node (SA Node): The SA node is a small group of specialized cells that is located high in the right upper atria. The SA node has the highest automaticity rate of any site in the heart and serves as the primary pacemaker, generating a heart rate of 60-100 beats per minute.

Atrioventricular Node (AV Node): The AV node is a small group of specialized cells located in the lower portion of the right atrium, near the tricuspid valve. The AV node has three major functions:

1. It slows the conduction of the impulse from the atria to the ventricles to allow the atria to fully depolarize and contract, allowing for atrial kick and adequate ventricular filling time.
2. The AV node has the second highest automaticity rate of 40-60 beats per minute, and functions as the backup pacemaker, should the SA node fail.
3. The AV node screens rapid atrial impulses to protect the ventricles from extremely fast rates.

Bundle of His: This is a short bundle of fibers at the bottom of the AV node leading down towards the bundle branches.

Bundle Branches: The bundle branches have the 3rd highest rate of automaticity and serve, as the tertiary backup pacemaker should the SA and AV node fail. The bundle branches are capable of generating a pacing rate of 20-40 BPM. The bundle branches are bundles of fibers located along the septum of the heart that convey electrical impulses to the right and left ventricles. The right bundle branch consists of one branch that supplies electrical impulses to the muscle cells of the right ventricle. The left bundle branch consists of two branches (anterior fascicle and posterior fascicle). The left ventricle has a much larger muscle mass than the right ventricle. To keep the ventricles in synch with each other, the left bundle branch consists of two fascicles that are utilized to quickly depolarize the larger muscle mass of the left ventricle, so that both the left and right ventricles contract at the same time. Without two fascicles the much larger left ventricle muscle mass would depolarize at a slower rate than the smaller muscle mass of the right ventricle. The end result would be asynchronous ventricular contraction.

Electrical Conduction Through The Heart

The SA-Node fires and produces an electrical impulse, which is conducted through the atria, causing them to depolarize, creating the P wave. The impulse is held in the AV-Node for 0.10 seconds, allowing the left and right atria to fully depolarize. The impulse is then conducted through the AV-Node to the Bundle of His and then to the left and right bundle branches. As the impulse leaves the AV-Node it picks up speed rapidly and conducts very quickly through the Bundle of His and left and right bundle branches, finally reaching the Purkinje fibers and ventricular myocardium.

Rates of Cardiac Pacing Sites:

SA-Node: 60-100 BPM
AV-Node: 40-60 BPM
Purkinje Fibers: 20-40 BPM

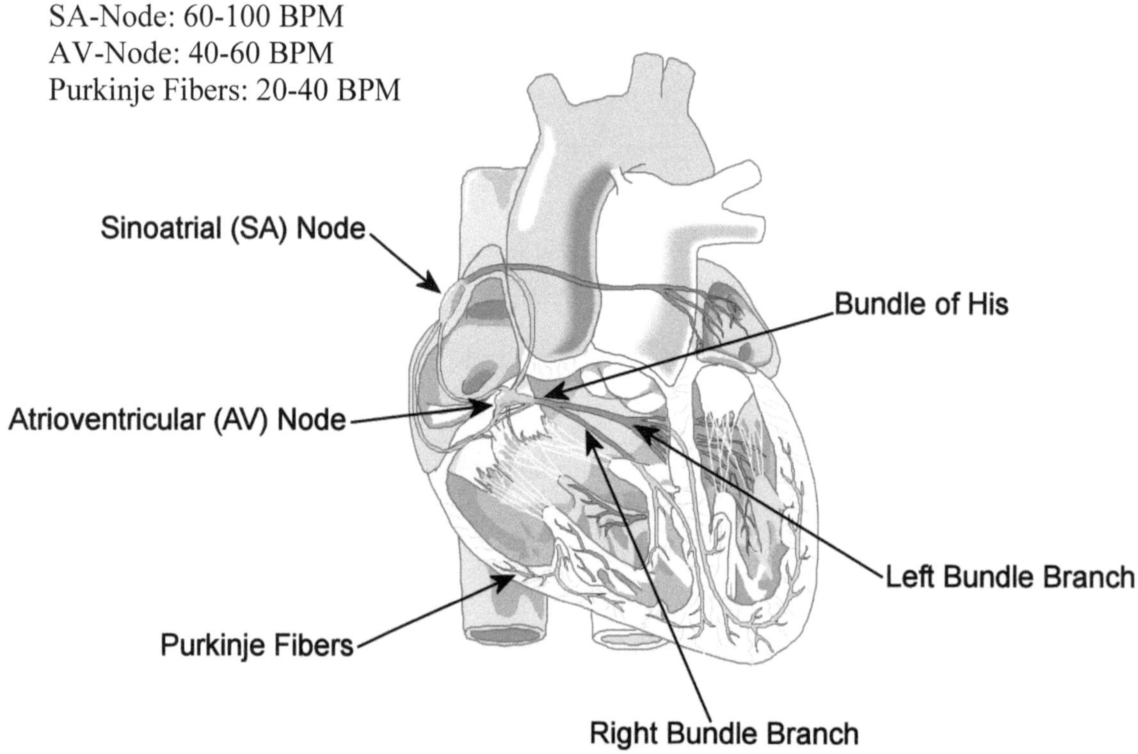

EKG Lead Basics

The EKG monitoring leads are typically of two types, 3 and 5 lead systems. As the name might imply, the 3 lead system consists of 3 electrodes and is capable of monitoring 3 views of the heart. The 5 lead system consists of 5 electrodes and is capable of monitoring 5 views of the heart.

The preferred standard monitoring lead is Lead II, and is the standard lead seen on a 12 lead EKG. However, Lead II is not considered the optimum monitoring lead. On a 3 lead system, lead MCL_1 is the preferred lead. On a 5 lead system the preferred lead is V_1.

Lead I produces a positive deflection on the EKG and provides a view of the heart that shows current moving from left to right. Lead I is useful for monitoring atrial rhythms and hemiblocks. Lead II produces a positive deflection and is useful for monitoring sinus and atrial activity. Lead III produces a positive deflection and is useful for monitoring changes associated with an inferior wall myocardial infarction.

The V_1 Lead shows the P wave, QRS and ST segment and is useful for monitoring ventricular rhythms and bundle branch blocks. The Modified Chest Lead or MCL_1, is used to monitor ventricular arrhythmias and to distinguish between ventricular tachycardia and supra-ventricular tachycardia. The MCL_1 lead is also useful in assessing bundle branch blocks, P wave changes and to confirm pacemaker wire placement.

Views of the heart from a 3 lead system.

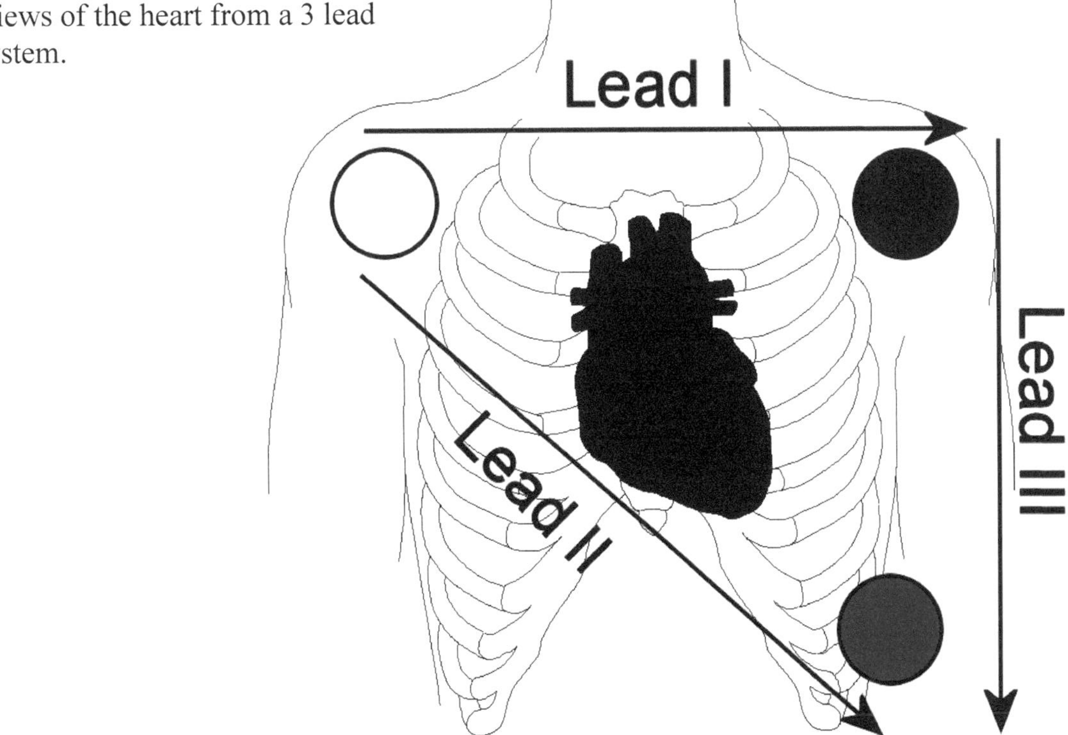

Placement of Electrodes on a 3 and 5 Lead System

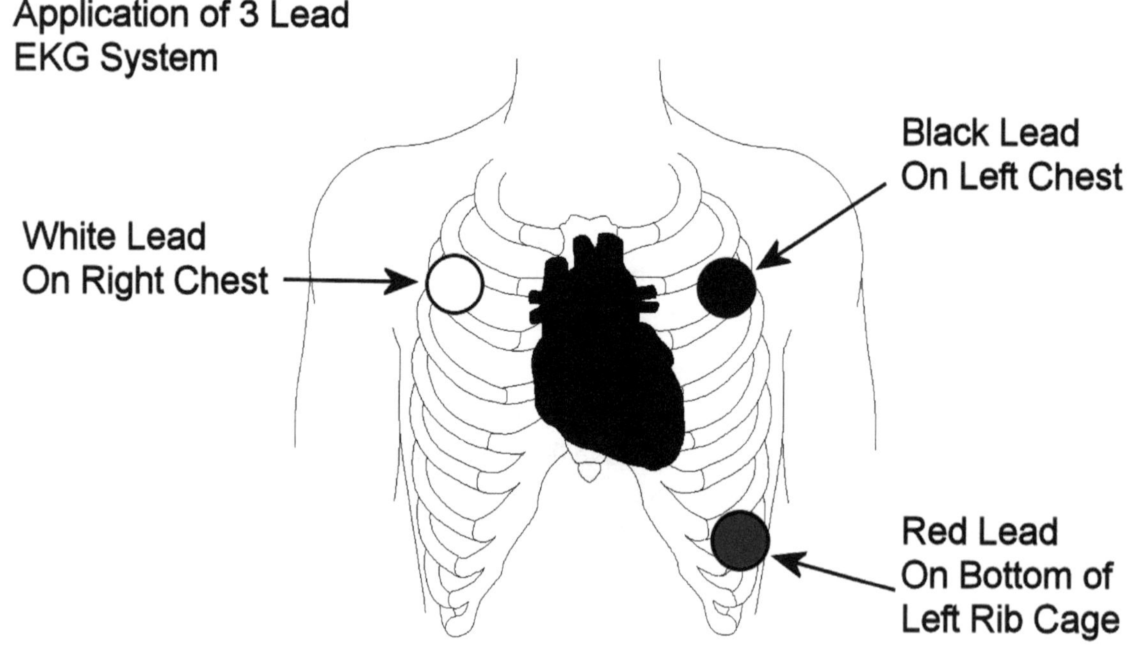

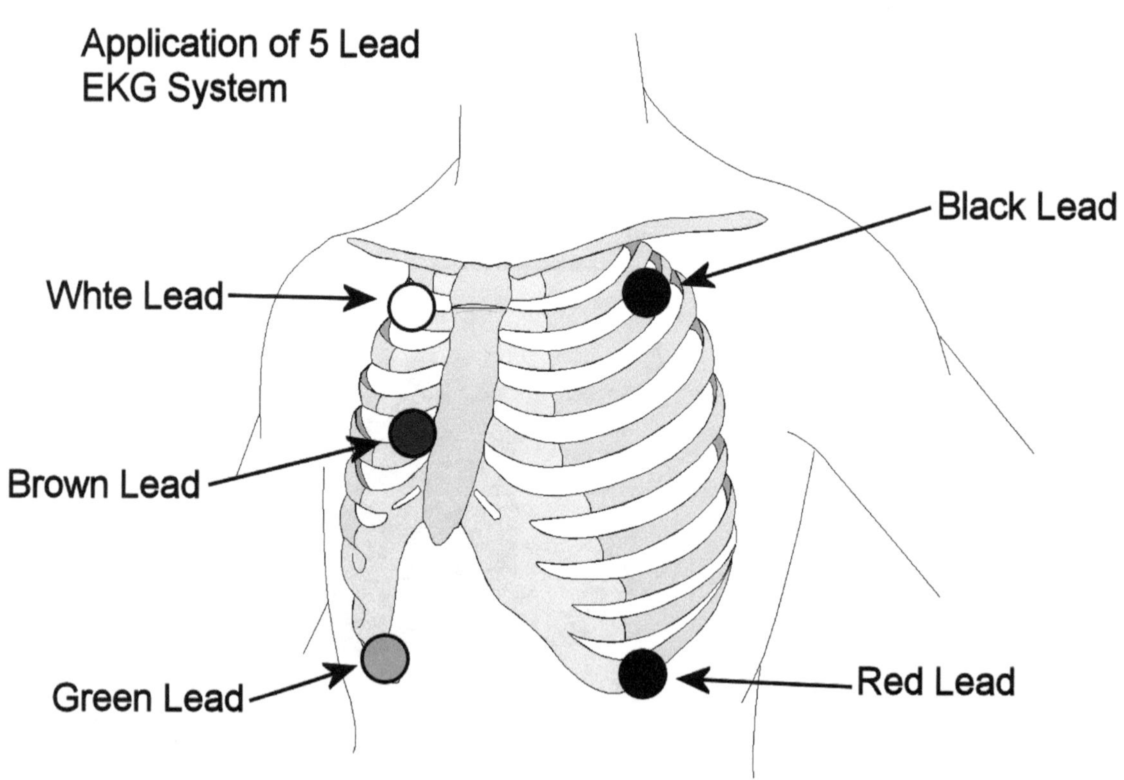

12 Lead Monitoring Using EASI™ Lead System

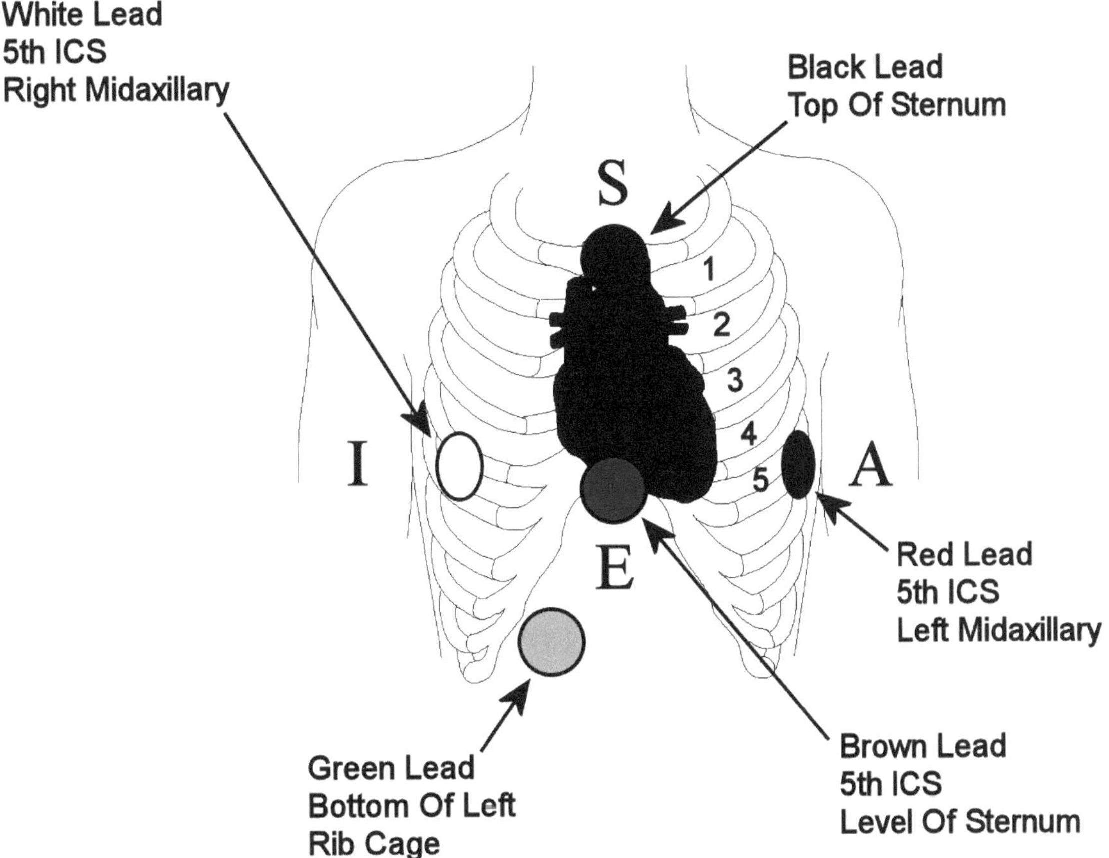

Placing Electrodes

When placing electrodes on a patient, it is important to choose an area over soft tissue, not over boney areas, adipose folds or large muscles. Doing so will contribute to large amounts of artifact being present in the EKG tracing.

Prepare the patients skin by thoroughly cleaning the area with soap and water and drying. Alcohol swabs can be used to remove skin oil if the area is relatively clean. If the patient has a fair amount of body hair, it would be beneficial to shave the areas that you will be applying the EKG patches. Not only will this provide better electrode to skin contact, but it will also save both you and your patient a lot of grief when the time comes to remove the electrodes.

Before applying the electrodes to the patients skin, attach the wires to the patches. It is unpleasant for the patient to have the electrodes place on the skin, and then a fair amount of pressure applied to get the EKG wires to snap onto the electrodes.

Troubleshooting

Electrode patches should be changed at least every 24 hours or more frequently if they are losing contact with your patient's skin.

Ensure that all wires are firmly snapped onto the electrode patches and firmly plugged into the EKG monitor. Lose wires will generate a large amount of artifact.
If you are having a difficult time obtaining a clean EKG tracing:

- Try new patches.
- Replace the EKG patches and wires.
- Relocate the EKG patches to get the best tracing. Sometimes this can be a trial and error process.
- Changing leads may give a cleaner EKG tracing.
- Some EKG monitors have a filter mode that can be turned on to filter out patient movement and muscle artifact.
- If the rhythm is paced, is the monitor accurately reading the heart rate? It may be counting the pacing spike as an *"R"* wave and reading double the actual heart rate. Check to see if your monitor has a special *"diagnostic mode"* that can be utilized to accurately read paced rhythms.

Basic EKG Rhythm Analysis Guide

When looking at a cardiac rhythm ask your self the following questions.

1. What is the rate? Is it fast or slow?

Rhythm	Rate
NSR	60-100
Sinus Brady	< 60
Sinus Tachycardia	100-150
SVT	>150
Junctional	40-60
Accelerated Junctional	60-100
Junctional Tachycardia	>100
Idioventricular Rhythm	20-40
Accelerated Idioventricular	40-100
Ventricular Tachycardia	>100

Calculating Heart Rates:

- For regular rhythms count the number of large boxes between R waves and divide into 300 or look up on the chart. Use the same method to calculate the atrial rate by counting large boxes between P waves.
- For irregular heart rates count the number of R waves in a 6 second strip and multiply by 10. Do the same for P waves to calculate atrial rate. Some rhythms may have atrial rates that differ from the ventricular rate.

# Of Large Boxes	Rate
1	300
2	150
3	100
4	75
5	60
6	50
7	43
8	37
9	33
10	30

2. Measure the following: PRI, QRS, and QT.

3. Analyze the relationship of the P wave and the QRS.

- Is there a P wave for each and every QRS Complex?
- Are the P waves upright and regular?
- Do the P waves march out regularly?

4. Analyze the QRS complex.

- Is the QRS narrow or wide? Grossly abnormal in appearance?
- If the QRS is narrow 0.12 or less, the impulse will have originated above the ventricles.
- If the QRS is very wide, 0.16 or greater with a T waves the opposite direction of the QRS complex. It would be suspected that the impulse is ventricular in origin.

5. Is the rhythm regular or irregular?
   - What rhythms are regular? Which rhythms are irregular?

6. Can you name the rhythm yet?

Putting It All Together

If there are consistently more P waves than QRS, you might consider the following:

- Premature Atrial Contractions, Second Degree AVB Type I and Type II.
- Complete Heart Block will have independent P wave rate and QRS rate. P waves are not conducted.

If the QRS complex is very wide, > 0.16 and no P waves consider:

- Idioventricular Rhythm
- Accelerated Idioventricular Rhythm
- Ventricular Tachycardia

If there are P waves for each and every QRS and the QRS complex is, narrow < 0.12 consider:

- Sinus Bradycardia
- NSR
- Sinus Tachycardia

If the QRS complex is narrow < 0.12 and the rate is > 150 consider:

- SVT (P waves may or may not be visible to due high rate).

If the P waves are absent or upside down with a QRS width < 0.12 and the rhythm is regular consider:

- Junctional Rhythm
- Accelerated Junctional Rhythm
- Junctional Tachycardia

If the P waves are absent and the QRS is < 0.12 and the rhythm is irregular consider:

- Atrial Flutter
- Atrial Fibrillation

If the PRI is > 0.20 with a P wave for every QRS and the QRS is < 0.12 consider:

- 1st AVB

If there is, no QRS present consider:

- Are the leads attached to the patient?
- Are the leads plugged into the monitor?
- Confirm absence of QRS in another lead.
- Does your patient have a pulse?

After checking all the above consider the following.

- Asystole
- Ventricular Fibrillation

ST Segment

Basics of the ST segment

The ST segment starts at the end of the QRS and ends at the start of the T Wave. The ST segment represents the early beginning of ventricular repolarization. The portion of the EKG tracing where the QRS ends and the ST segment begins is called the *J Point*.

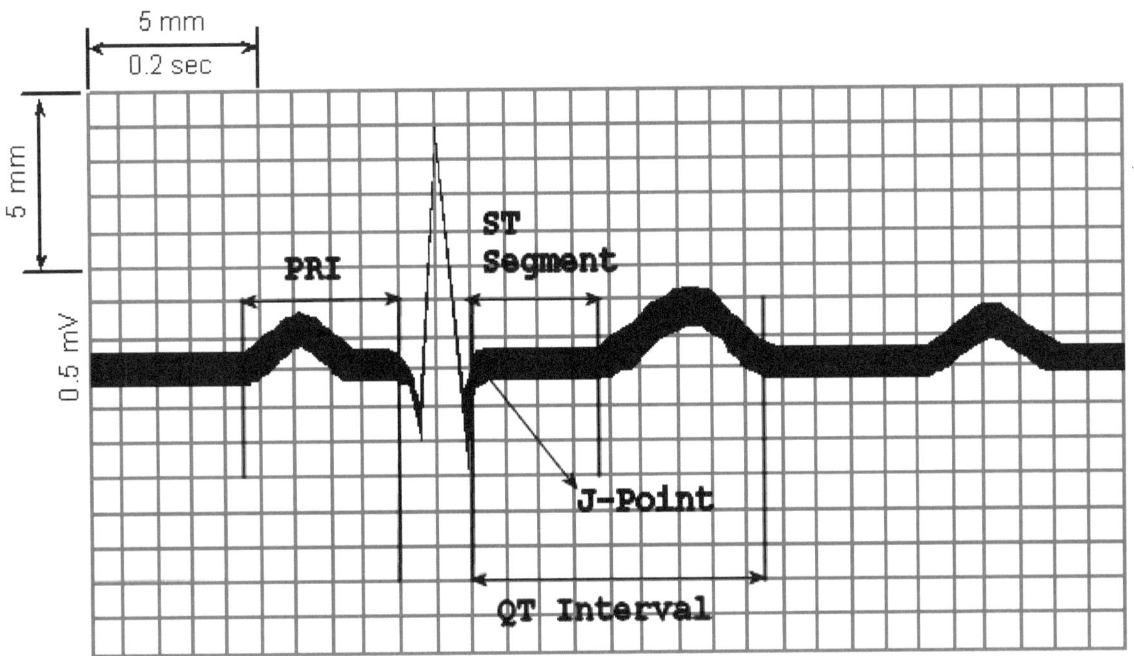

Characteristics: The ST-segment normally remains *"Isoelectric"* which is the normal baseline of the EKG. An abnormal ST segment is caused by abnormal ventricular repolarization. Elevation greater than 1mm in two or more reciprocal leads may indicate injury. Depression greater than 1mm in two or more reciprocal leads may indicate myocardial ischemia. A 12-lead EKG is indicated for further evaluation.

Drawing a *"Cardiac Enzyme Panel"* may also reveal further information about that patient's cardiac status. A *"Cardiac Enzyme Panel"* is a generic term to refer to standard set of labs such as "CK, CK-MB, Troponin-T Index and Troponin -I Index". If the patient has a known history of CHF, a BNP panel may also be indicated to gauge severity of CHF.

Possible causes of ST segment abnormalities:

Causes of ST Depression	Causes of ST Elevation
Artifact	Acute myocardial Infarction
Bundle Branch Block	Acute Pulmonary Embolism: In a 12 lead EKG leads, V_1 and AVR may reflect a dilated right heart. Lead V_1 associated with RBBB and positive T Waves may raise the question of an Acute PE.
Digitalis: "Scooped out"	Artifact
Hypokalemia	Benign Early Repolarization / J Point elevation
Hypothermia	Burgada Syndrome
Myocardial ischemia	Electrolyte Imbalances such as hyperkalemia.
Reciprocal EKG Changes	Left Bundle Branch Block
Subendocardial Infarct	Medications
Tachycardia	Pacemaker with paced beats from the right ventricle.
Ventricular Hypertrophy	Pericarditis
	Prinzmetal Angina (Variant Angina): Chest pain with elevated ST segment that returns to normal in minutes.
	Subarachnoid hemorrhage
	Transthoracic Cardioversion
	Ventricular Aneurysm
	Ventricular hypertrophy

Examples of ST Segment Changes

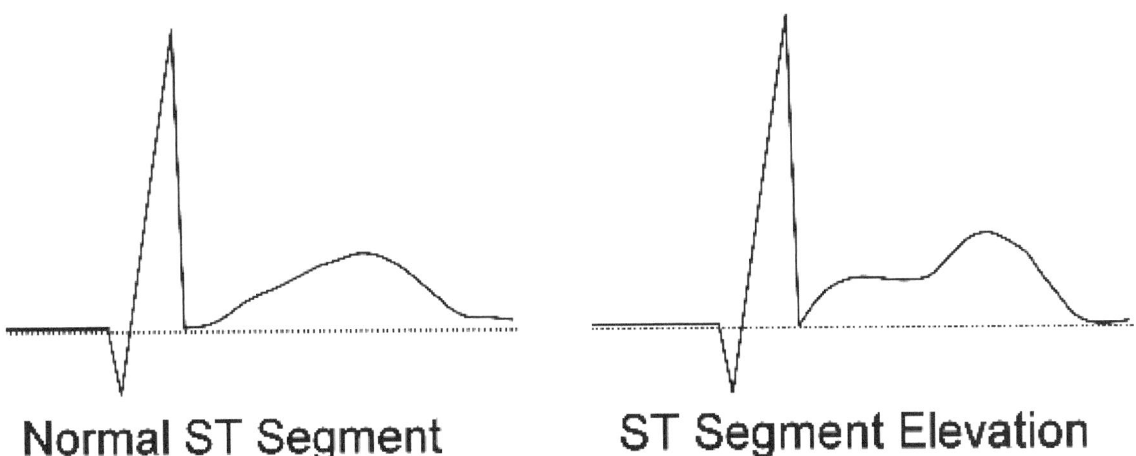

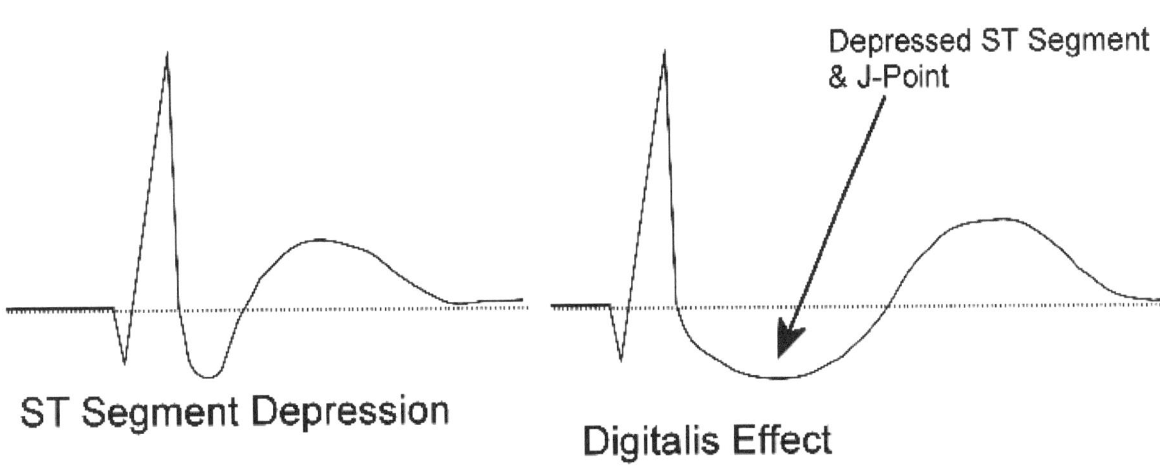

Electrolyte Imbalances:

Hyperkalemia: The ST segment may not be visible when the serum K+ level is a value of 6 mEq/L.
Hypocalcemia: The ST segments may become prolonged.

Drugs:

Quinidine and procainamide: both of these drugs have been known to cause 1mm or greater ST depression. Consider this as a potential cause if you have a patient on one of these drugs.

Ventricular Hypertrophy:

With right ventricular hypertrophy, the ST segment may be depressed 1mm or more in the following leads: II, III, aVF, V_1, V_2, V_3.
With left ventricular hypertrophy, the ST segment may be depressed in the following leads: I, aVL, and V_5-V_6.

Pericarditis:

Area of Pericarditis	Leads of ST segment elevation
Generalized	I, II, III, aVL, aVF, V_2-V_6
Anterior	I, V_2-V_4
Lateral	I, aVL, V_5-V_6
Inferior	II, III, aVL

QT Interval

Definition: The QT interval (QTI) represents total ventricular activity. It begins with the first wave in the *QRS complex* representing ventricular depolarization, and ends when the *T wave* returns to baseline at the isoelectric line, representing ventricular repolarization.

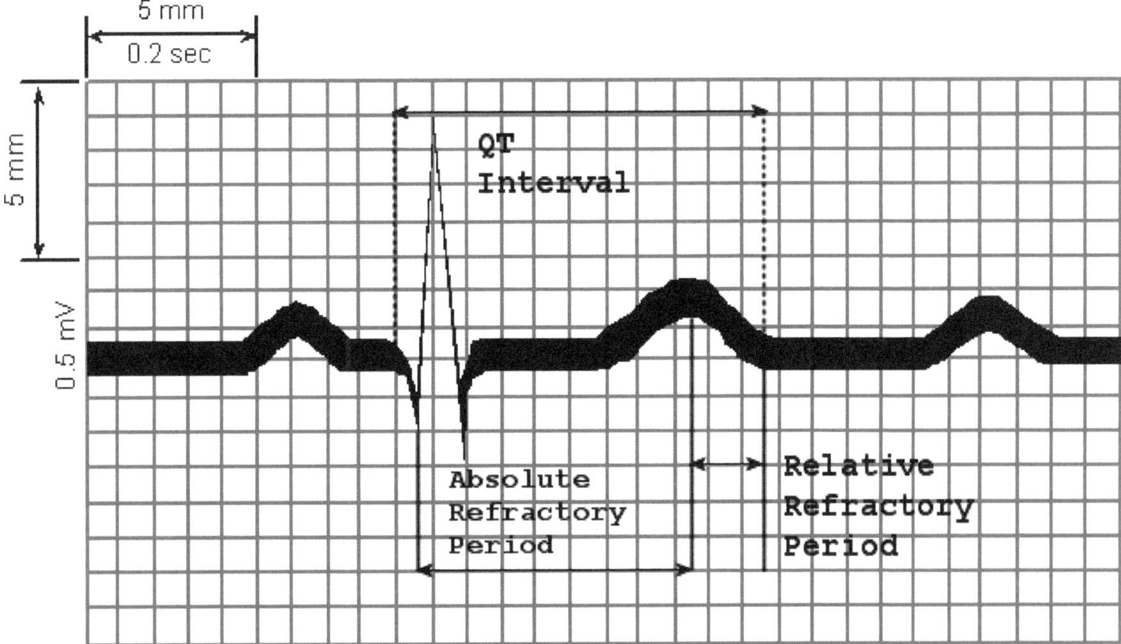

Clinical significance of QT interval: A prolonged QT represents a slowing in ventricular repolarization. This may be due to medications or simply a slower heart rate. The length of the QTI is directly related to heart rate. A slower heart rate, such as sinus bradycardia will have a longer QTI than sinus tachycardia, which will have a shorter QTI. This variation in QTI length is directly related to the speed of depolarization and repolarization of the ventricular muscle mass. A QTI that is prolonged will have a increased relative refractory period, which places the patient at a increased risk of the possibility of the ventricles responding to a ectopic beat, such a R on T event causing a lethal dysrhythmia.

Normal Measurements of the QTI: 0.36-0.40
It is best to measure the QTI in a lead in which the T wave is most pronounced. The normal overall length of the QTI should be equal to or less than half of the R-R interval.

Causes of abnormal QTI measurements:

Prolonged QTI	Shortened QTI
Acute myocarditis	Any rapid heart rate.
AMI	Digitalis
Amiodarone	Hypercalcemia
Arrythromycin based antibiotics	Hyperthermia
Droperidol	Supraventricular tachycardia
Hypocalcemia	
Hypokalemia	
Hypothermia	
Large doses of haloperidol	
Large Doses of prochlorperazine	
Left ventricular hypertrophy	
Pericarditis	
Procainamide	
Quinidine	

Large dosages of antiarrhythmic drugs or the combination of antiarrhythmics may induce lethal heart rhythms know as "Pro-arrhythmias" such as ventricular tachycardia or torsade de pointes. You are strongly advised to consult your ACLS provider manual for information on the use antiarrhythmics and their proper administration.

Normal Sinus Rhythm

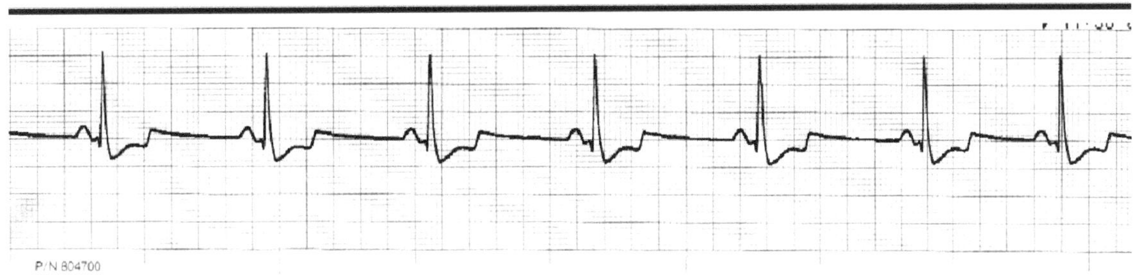

Rate: 60-100 Beats Per Minute (BMP)
PRI: 0.12-0.20 sec
QRS: < 0.12 sec
QT: < 0.40 sec
Rhythm: Regular
Source of pacer: SA Node

Characteristics: Primary rhythm typical of most adults. Focus of care, should be on the patient's overall condition, and vital signs. Changes from this rhythm to other arrhythmias should be a cause of concern.

Treatment: None.

Sinus Bradycardia

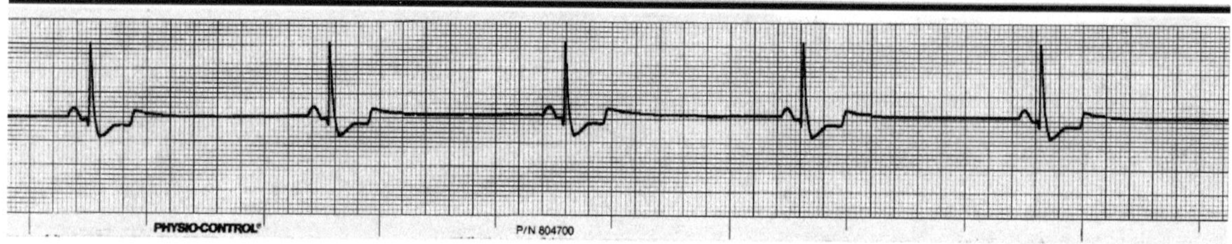

Rate: < 60 BPM
PRI: 0.12-0.20 sec
QRS: < 0.12 sec
QT: < 0.40 sec. May be prolonged with excessively low heart rates.
Rhythm: Regular
Source of pacer: SA Node

Characteristics: This rhythm may be normal for well-conditioned athletes and during sleep. Sinus bradycardia, may be caused by several factors such as increased vagal tone from vomiting, bearing down to have a bowel movement or from medications such as digitalis, calcium channel blockers, beta blockers and many other antiarrhythmic medications; common with inferior wall MI, obstructive jaundice and increased intracranial pressure (ICP).

Treatment: Treatment is only necessary if the patient is symptomatic. Atropine 0.5 mg, to a maximum of 3.0 mg. Consider external transcutaneous pacing. Be prepared to assist the physician with the placement of an external temporary pacer. Treatment of associated hypotension may also need to be addressed.

Sinus Tachycardia

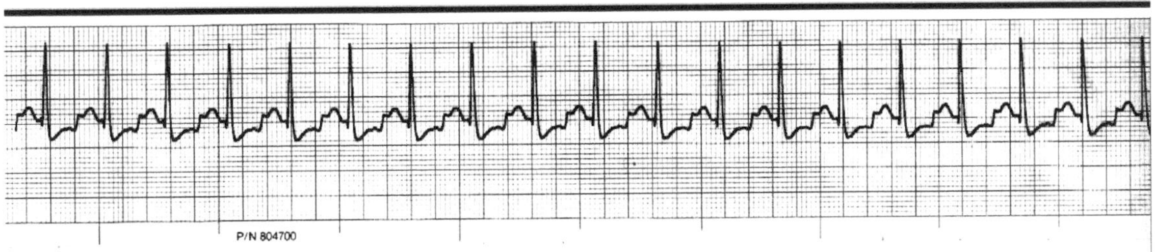

Rate: 100-150 BPM
PRI: 0.12-0.20 sec
QRS: < 0.12 sec
QT: < 0.40 sec
Rhythm: Regular
Source of pacer: SA Node

Characteristics: Sinus tachycardia is a normal response to stress and exercise. If it is persistent, at rest it may indicate a more severe underlying problem such as fever, dehydration, blood loss, anemia, anxiety, heart failure, hypermetabolic states or ingestion of a significant stimulant such as cocaine or methamphetamine. Drugs that can cause sinus tachycardia are atropine, isoproterenol, epinephrine, dopamine, dobutamine, norepinephrine, nitroprusside and caffeine. Sinus tachycardia increases the hearts need for oxygen, decreases ventricular diastolic time and decreases coronary artery perfusion. Reflexive sinus tachycardia is often seen in hypotensive patients, in an attempt to maintain adequate blood pressure.

Treatment: The underlying cause must be identified and treated. Drugs that may be given to slow the heart are: digitalis, beta-blockers, calcium-channel blockers, sedatives and various other antiarrhythmics medications.

Sinus Arrhythmia

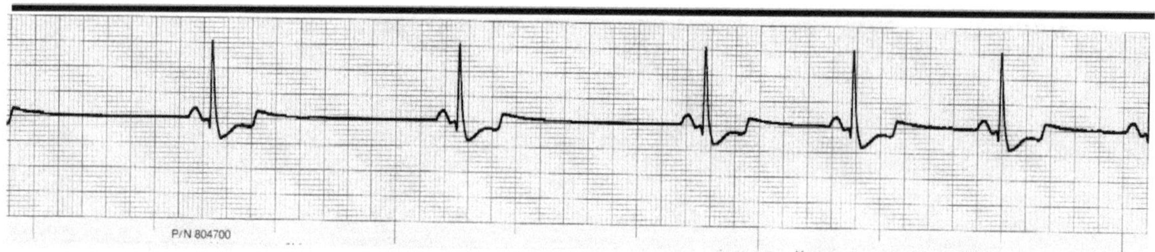

Rate: Variable
PRI: 0.12-0.20 sec
QRS: < 0.12 sec
QT: < 0.40 sec
Rhythm: Irregular
Source of pacer: SA Node

Characteristics: The pacemaker cells of the SA node fire at an uneven rate causing an irregular rhythm. The irregularity of the rhythm corresponds with the respiratory cycle. The heart rate will increase with inspiration and decrease with expiration. During inspiration, the blood return back to the heart reduces vagal tone and results in an increased heart rate. During expiration, venous return to the heart decreases, resulting in increased vagal tone and a decrease in heart rate.

Sinus arrhythmia can occur in athletes, children and sometimes in older adults. Sinus arrhythmia may occur in conditions that are unrelated to the respiratory cycle such as: inferior wall MI, advanced age, use of morphine or digitalis, CAD and changes in ICP. To clarify the underlying rhythm, ask your patient to hold his/her breath for a few seconds. The cardiac rhythm should revert to a regular sinus rhythm.

Treatment: Benign condition, no treatment is needed.

Premature Atrial Contractions (PAC)

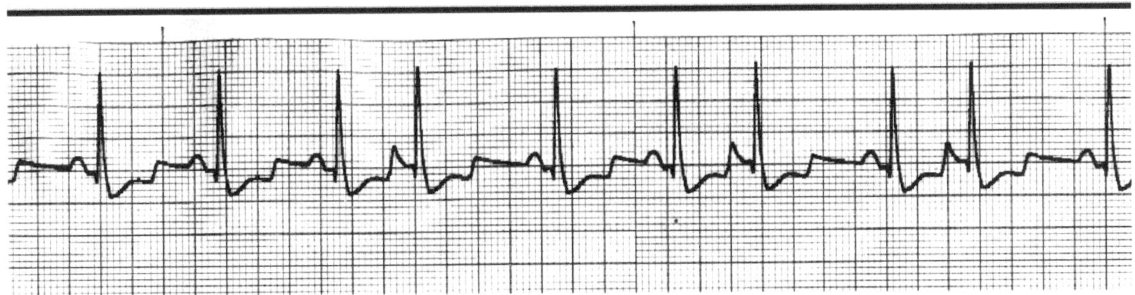

Rate: Variable
PRI: 0.12-0.20 sec
QRS: < 0.12 sec
QT: < 0.40 sec
Rhythm: Irregular when extra beats are present.
Source of pacer: Irritable focus in the atria fires before the next sinus impulse is due, resulting in an early beat.

Characteristics: PAC's usually have a noncompensatory pause, which is the interval following the PAC, (look the left of the pause for the PAC). Premature depolarization of the SA node causes it to reset itself and attempt to start a regular rhythm. The P wave may vary in shape and size and may appear to be abnormal. This is in part due to the location of the irritable atrial focus that originated the impulse. PACs occur in patients with COPD, CHF, MI, anxiety, hypermetabolic states and from ingestion of caffeine, nicotine and ETOH. Emotions, infections, electrolyte imbalances, digitalis toxicity, drugs such as isoproterenol, norepinephrine and epinephrine may also cause PAC's.

Treatment: PAC's are not dangerous by themselves. However, they can serve as a warning sign for the development of more clinically significant arrhythmias. PACs may be treated with digitalis, beta-blockers and calcium channel blockers. Obtaining digitalis levels and a urine drug screen may be warranted along with possibly finding and treating the underlying cause.

Supraventrical Tachycardia

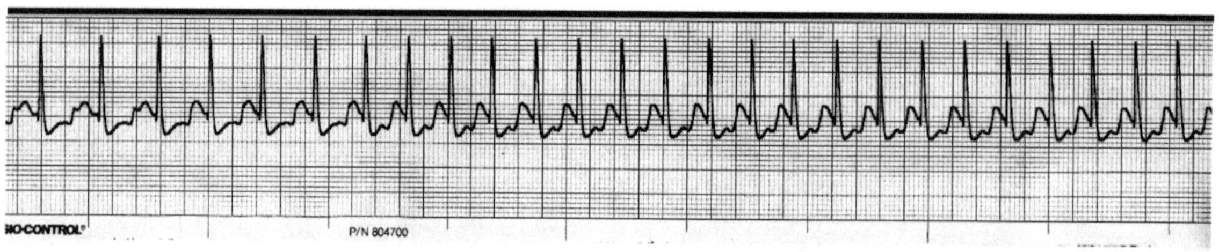

(PSVT/PSAT)

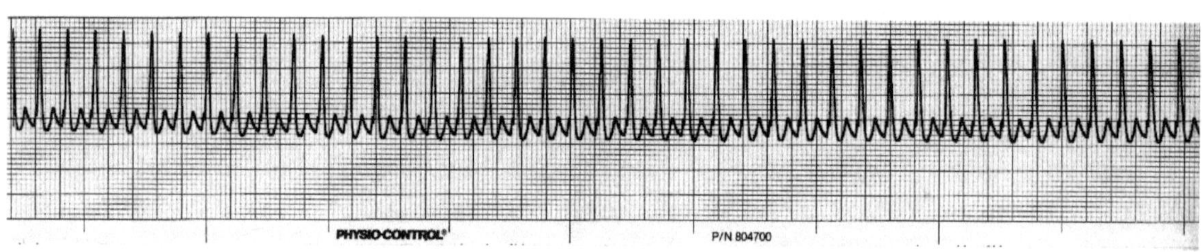

(SVT)

Rate: > 150 BPM
PRI: Usually < 0.12 sec
QRS: Narrow, < 0.12 sec
QT: < 0.40 sec
Rhythm: Regular
Source of pacer: Atrial re-entry current

Characteristics: Tachycardias with a narrow QRS < 0.12 seconds and faster than 150 BPM do not originate from the SA node, since the upper limit rate for the SA node is 150. The rhythms can vary in name from supraventricular tachycardia (SVT), atrial tachycardia, or a rhythm that may speed up and slow down called paroxysmal atrial tachycardia (PAT) or paroxysmal supraventricular tachycardia (PSVT). If P waves are visible, the rhythm may be called atrial tachycardia, if no P waves are visible due to a very fast rate, label the rhythm SVT.

The onset of PSVT/PAT is typically abrupt, with the onset being initiated by a premature atrial beat. The arrhythmia may self terminate in a few minutes to a few hours.

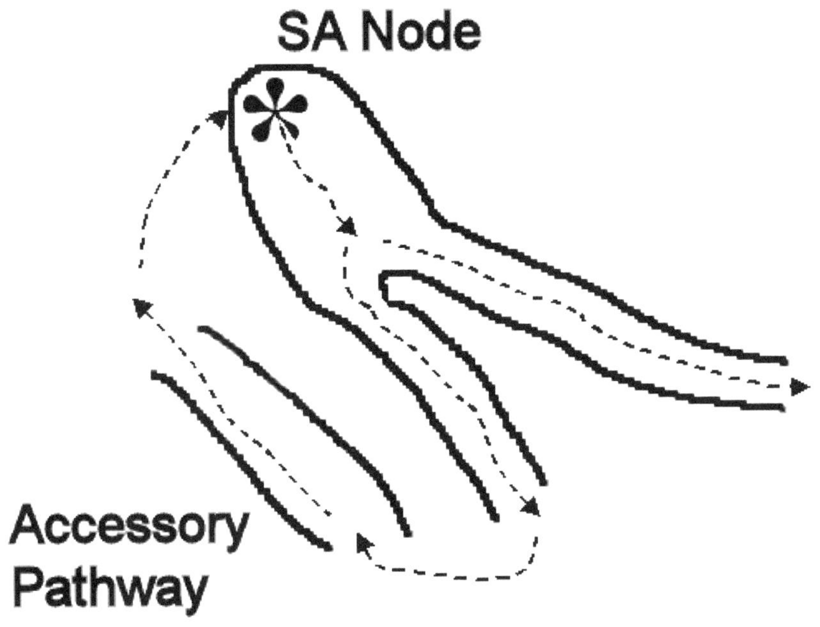

Both SVT and PSVT may involve AV nodal reentry or the use of an accessory conduction pathway between the atria and ventricles. The electrical impulse returns to the SA node or atria and results in stimulation of an ectopic beat. This process continues resulting in an extremely fast heart rate.

Many of the causes that trigger SVT/PSVT are the same that cause PACs. Please see this section for possible causes.

Treatment: It is worth noting that this rhythm can be extremely detrimental to patients suffering from a MI, since it increase the myocardium's consumption of oxygen and results in further extending the infarction. Since the heart is beating extremely fast, the ventricular diastolic filling time is insufficient to allow for adequate filling of the ventricles. The result is a drop in cardiac output and blood pressure. If the patient is unstable, emergent cardioversion is the priority. Vagal maneuvers may be performed by a physician. A stable patient may be treated with beta-blockers, calcium channel blockers, digoxin, amiodarone or adenosine.

Atrial Flutter

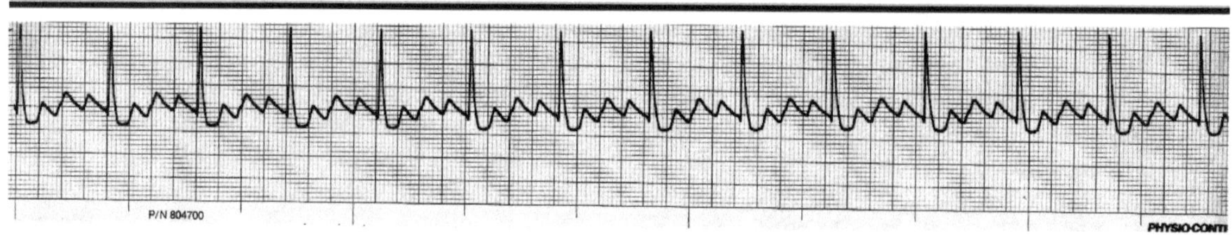

Rate: Irregular
PRI: None
QRS: < 0.12 sec
QT: < 0.40 sec
Rhythm: Irregular to regular
Source of pacer: Atrial

Characteristics: In atrial flutter, P waves are not present, since the atria are moving at such a fast rate. They form a "Saw Tooth" pattern called F waves. Typically, one F wave is hidden in the QRS or T wave. In the strip above, only 3 F waves are seen as the 4th is hidden in the QRS. This is a 4:1 atrial flutter. The cause of the arrhythmia is possibly due to reentry current within the atria, or a rapid firing ectopic focus in the atria. Because of the rapid atrial rate, the AV node blocks at least every other F wave and often blocks 2 or 3 F waves in a row, in order to protect the ventricles from the rapid rate.

Atrial flutter leads to loss of atrial contraction and "Atrial Kick" resulting in decreased cardiac output by 20-30%. This may result in the patient becoming hemodynamically unstable. In addition, the patient is at risk for forming mural thrombi, resulting in systemic or pulmonary embolism.

Causes: MI, rheumatic heart disease, thyrotoxicosis, CHF and ischemia.

Treatment: Treatment may very. If this is an acute arrhythmia, the patient may be cardioverted, or treated with beta-blockers, calcium channel blockers, digoxin, amiodarone or procainamide. If this is a chronic rhythm that would not convert with cardioversion or medications, it is important that the patient be evaluated and possibly placed on anticoagulation medication before being discharged home.

Atrial Fibrillation

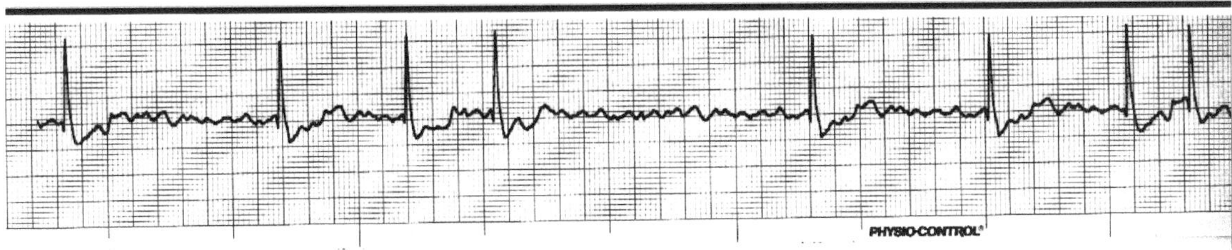

Rate: Variable
PRI: None
QRS: < 0.12 sec
QT: < 0.40 sec
Rhythm: Irregular
Source of pacer: Atrial

Characteristics: Atrial fibrillation is caused by chaotic ectopic or reentry current activity, which causes the atria to quiver rather than contract. The atria quiver at a high rate producing the fuzzy and garbled waveforms seen where a flat isoelectric line should be. Atrial fibrillation can generate a ventricular response rate that is controlled or very fast and can place the patient at risk for hemodynamic instability.

Cardiac output is reduced with the loss of "Atrial Kick" since the atria are not contracting. The ventricular rate my also be very fast resulting in further decreased cardiac output. In addition, since the atria are not contracting, the patient is at risk for the formation, of emboli leading to pulmonary embolism or stroke.

Causes: MI, rheumatic heart disease, COPD, CHF, ischemia chest trauma, CAD and open-heart surgery.

Treatment: Treatment may very. If this is an acute arrhythmia, the patient may be cardioverted, or treated with beta-blockers, calcium channel blockers, digoxin, amiodarone or procainamide. If this is a chronic rhythm that would not convert with cardioversion or medications, it is important that the patient be evaluated and possibly placed on anticoagulation medication before being discharged home.

Junctional Rhythm

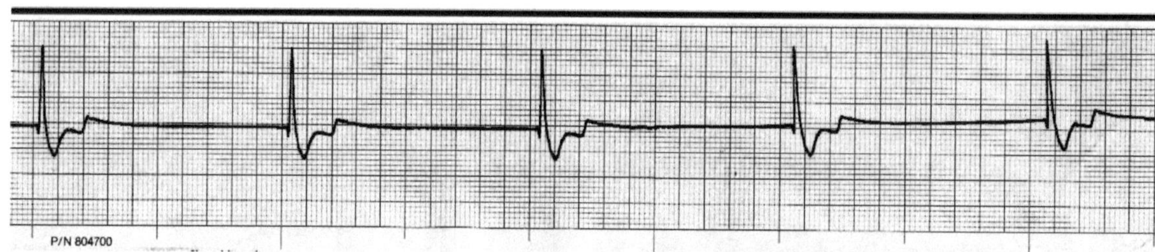

Rate: 40-60 BPM
PRI: Absent or variable
QRS: Usually < 0.12 sec. May be wider depending on pacer site.
QT: < 0.40 sec. May be prolonged with slower heart rates
Rhythm: Regular
Source of pacer: AV Node/Junction/Bundle of HIS

Characteristics: Rhythms that originate from the AV Junction: Cells around the AV node have automaticity and are capable of becoming the primary pacemaker if the SA node should fail. Junctional escape beats occur when the SA node fails to initiate an impulse or its rate falls below the rate of the AV node. Junctional rhythms can accelerate above the rate of SA node, and become the primary pacemaker and assume control of the heart's rhythm and rate. Atrial kick may be lost resulting in decreased cardiac output.

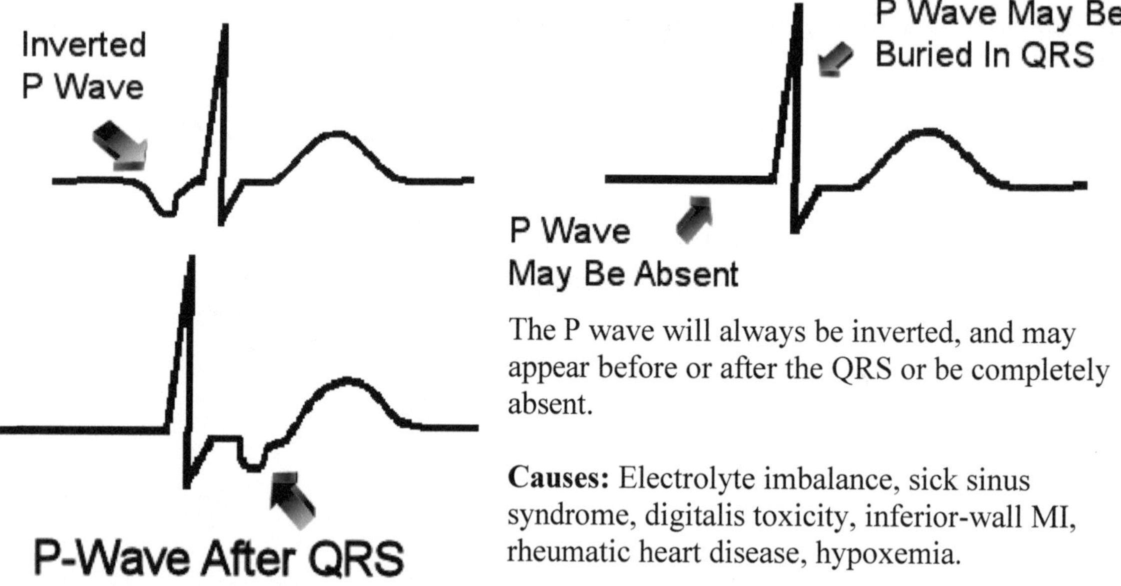

The P wave will always be inverted, and may appear before or after the QRS or be completely absent.

Causes: Electrolyte imbalance, sick sinus syndrome, digitalis toxicity, inferior-wall MI, rheumatic heart disease, hypoxemia.

Treatment: Find and treat reversible causes, temporary or permanent pacer, atropine 0.5 – 1.0 mg may cause the SA node to overdrive the AV node and increase the heart rate. Treatment is only needed if the patient is hypotensive or presents with hemodynamic instability.

Accelerated Junctional Rhythm

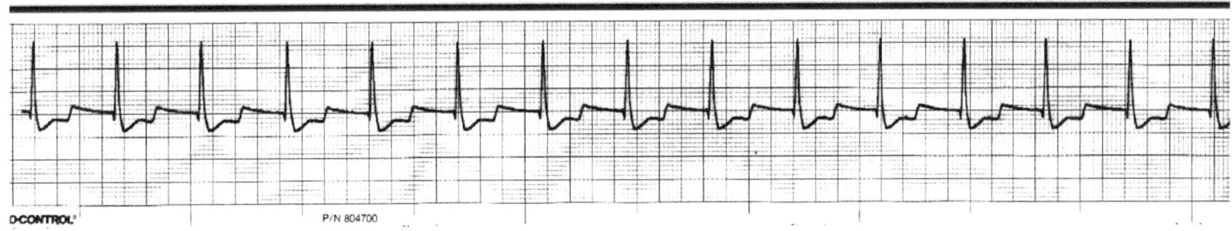

Rate: 60-100 BPM
PRI: Absent or variable
QRS: Usually < 0.12 sec. May be wider depending on pacer site.
QT: < 0.40 sec
Rhythm: Regular
Source of pacer: AV Node/Junction/Bundle of HIS

Characteristics: Same as a junctional rhythm. A junctional rhythm rate is typically 40-60. An accelerated junctional rhythm is a junctional rhythm with a rate of 60-100. P waves will be inverted, and can appear before or after the QRS or be completely absent. Atrial kick may be lost resulting in decreased cardiac output.

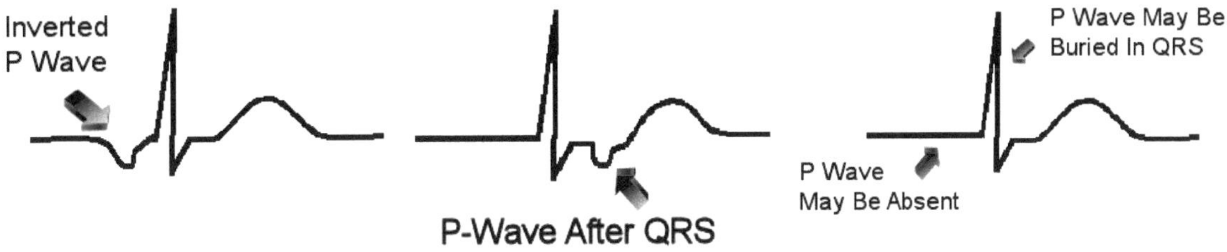

Treatment: Find and treat reversible causes, temporary or permanent pacer, atropine 0.5 – 1.0 mg may cause the SA node to overdrive the AV node and increase the heart rate. Treatment is only needed if the patient is hypotensive or presents with hemodynamic instability.

Junctional Tachycardia

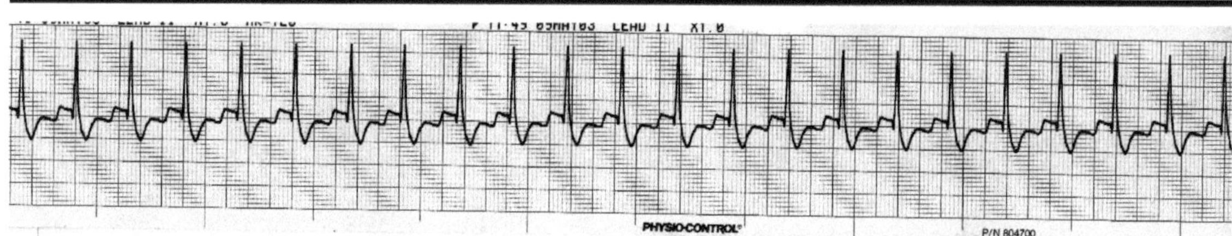

Rate: > 100 BPM
PRI: Absent or variable
QRS: Usually < 0.12 sec. May be wider depending on pacer site.
QT: < 0.40 sec
Rhythm: Regular
Source of pacer: AV Node/Junction/Bundle of HIS

Characteristics: Same as a junctional rhythm. A junctional rhythm rate is typically 40-60. Junctional Tachycardia is a junctional rhythm with a rate of 100-150. P waves will be inverted, and can appear before or after the QRS or be completely absent. Atrial kick may be lost resulting in decreased cardiac output.

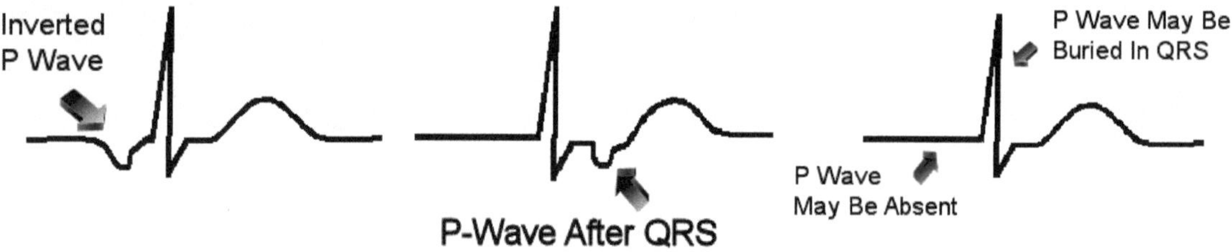

Treatment: Find and treat reversible causes, drugs such as beta-blockers, calcium channel blockers may slow or terminate the rhythm. Junctional Tachycardia is a common sign of digitalis toxicity.

Premature Junctional Contraction

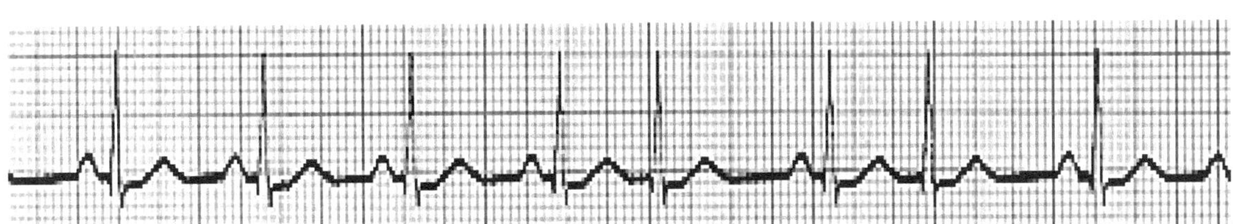

Rate: That of underlying rhythm
PRI: 0.12-0.20 sec
QRS: < 0.12 sec
QT: < 0.40 sec
Rhythm: Irregular
Source of pacer: Ectopic source in the AV Node/Junction/Bundle of HIS

Characteristics: The P wave will be inverted, and may appear before or after the QRS or be absent. A PJC is an early beat that originates from the AV junction. There will be a compensatory pause as the SA node resets. Look to the left of the pause, and you will most likely be able to locate the PJC. PJCs may occur singly or in groups. In groups of three or more, junctional tachycardia should be considered.

Frequent PJCs of more than six per minute may indicate enhanced automaticity or a reentry pathway in the AV junction. This may lead to more serious junctional arrhythmias.

Causes: Digitalis toxicity, enhanced automaticity of AV node, increased vagal tone, excessive dose of cardiac drugs, hypoxia, CHF, AMI and CAD.

Treatment: Find and treat reversible causes, drugs such as beta-blockers, calcium channel blockers may slow or terminate the rhythm.

1st Degree AV Block

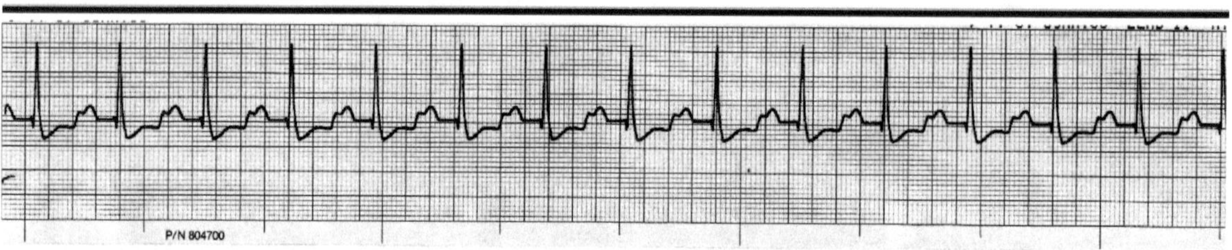

Rate: Variable
PRI: > 0.20 sec
QRS: < 0.12 sec
QT: < 0.40 sec. May be longer with slower heart rates.
Rhythm: Regular
Source of pacer: SA Node

Characteristics: 1st degree AV block, is characterized by what appears to be a normal sinus rhythm. Closer examination of the PRI will show that it is longer than 0.20 seconds. The PRI should not vary from one beat to the next. 1st degree AV block represents a delay in the conduction of electrical impulses through the AV node, which under normal circum stances would only delay the impulse 0.10 seconds or less. 1st degree AV block may progress into higher degrees of block.

Causes: Acute inferior MI, right ventricular infarction, increased vagal tone, ischemic heart disease, digitalis toxicity, beta-blockers, amiodarone, calcium channel blockers, electrolyte imbalances, rheumatic heart disease or myocarditis.

Treatment: Find and treat causative agent, observe for progression to higher blocks. Be prepared to pace, if the patient is bradycardic and symptomatic.

2nd Degree AV Block (Mobitz Type I, Wenkebach)

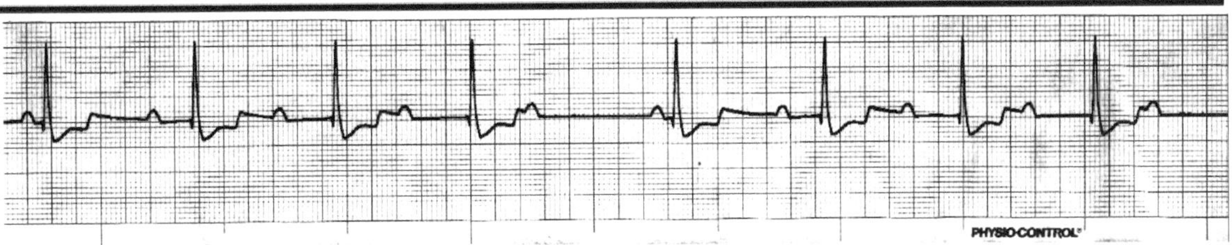

Rate: Variable
PRI: Progressive elongation of PRI until a P wave is not conducted.
QRS: < 0.12 sec
QT: < 0.40 sec. May be longer with slower heart rates.
Rhythm: Irregular due to dropped P waves.
Source of pacer: SA Node

Characteristics: This is a progressive slowing in the conduction from the atria to the ventricles until a beat is dropped. 2nd degree AV block (Wenkebach) appears on the rhythm strip as a progressively longer PRI on conducted beats until the impulse is not conducted through the AV node. This is denoted by a P wave (non-conducted) that is not followed by a QRS complex. The rhythm is repetitive in nature. 2nd degree AV block type I can be transient and reversible. However, it can progress to a higher degree of AV block.

Causes: Acute inferior MI, right ventricular infarction, increased vagal tone, ischemic heart disease, digitalis toxicity, beta-blockers, amiodarone, calcium channel blockers, electrolyte imbalances, rheumatic heart disease or myocarditis.

Treatment: Most of the time this rhythm produces no signs and symptoms and requires no treatment. If needed 2nd degree AVB Type I will respond to atropine if the patient becomes bradycardic and hypotensive. Temporary pacing should also be considered. Find and treat reversible causes, and observe for progression into higher forms of block.

2nd Degree AV Block (Mobitz Type II)

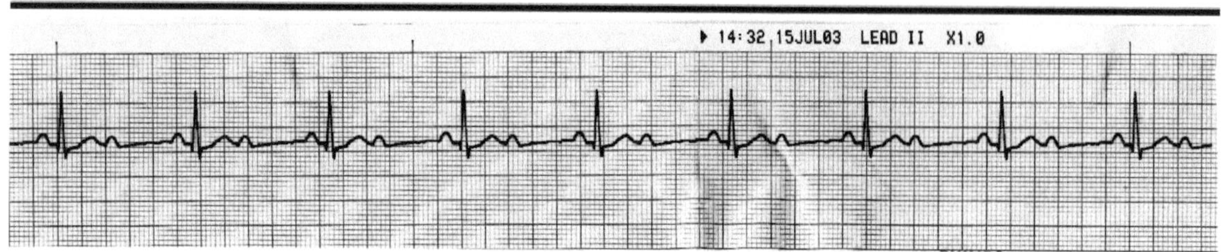

Rate: Variable
PRI: 0.12-0.20 sec when conducted
QRS: > 0.12 sec. May be longer depending on where the location of the block and if a bundle branch block is present.
QT: < 0.40 sec. May be longer with slower heart rates.
Rhythm: Irregular due to blocked beats.
Source of pacer: SA Node

Characteristics: 2nd Degree AVB Type II is a sudden failure of the conduction of a SA node impulse <u>without</u> a progressive elongation of the PRI segment of conducted P waves. This type of block is caused by the complete block of the impulse in a bundle branch and an intermittent block in the other bundle branch. This rhythm is characterized by more P waves than QRS complexes and a normal PRI when an impulse is conducted. The QRS complex is typically abnormal (wider than 0.12 seconds) due to the bundle branch block. The block may be in a ratio of 2:1 (two P waves for every QRS), 4:3, and 3:2.

Causes: Can be caused by damage to the bundle branch system following an acute anterior AMI. This is not caused by medications or increased vagal tone.

Treatment: This rhythm should be considered more serious than 2nd degree AVB type I. Type II can progress into complete heart block, or even ventricular asystole. A temporary pacemaker may have to be used if the patient becomes bradycardic and hypotensive. Atropine is not recommended.

3rd Degree AB Block / Complete Heart Block

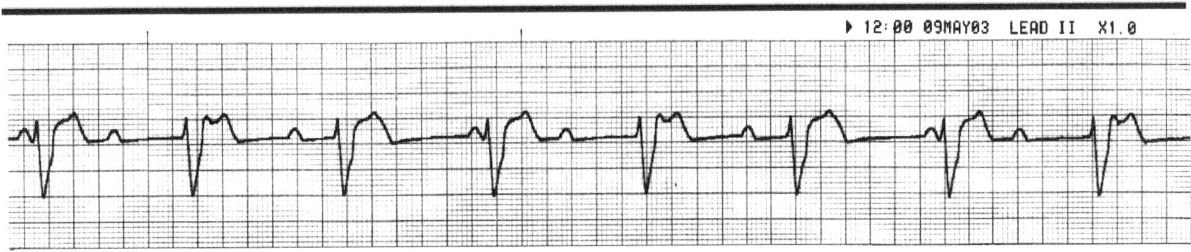

Rate: Depends on site of pacer.
PRI: No relationship between P waves and QRS complex.
QRS: Normal if junctional rhythm, wider if ventricular rhythm.
QT: May be < 0.40 seconds if it is a junctional rhythm or may be wider (> 0.40 sec) with slower ventricular rhythms.
Rhythm: Usually regular
Source of pacer: Variable and will be below the atria, since the connection between the SA node and AV node has been severed. The pacer site is usually Junctional or Ventricular in origin.

Characteristics: The connection between the atria and the ventricles has been severed. The resulting rhythm will be either junctional or ventricular in origin. It is important to march the P waves and QRS complexes out to establish that there is truly no relationship between the two. Since there is no connection between the atria and ventricles, it is possible to have an atrial rhythm such as atrial fibrillation or flutter, and a junctional or idioventricular rhythm driving the ventricles.
This is illustrated in the EKG strip below.

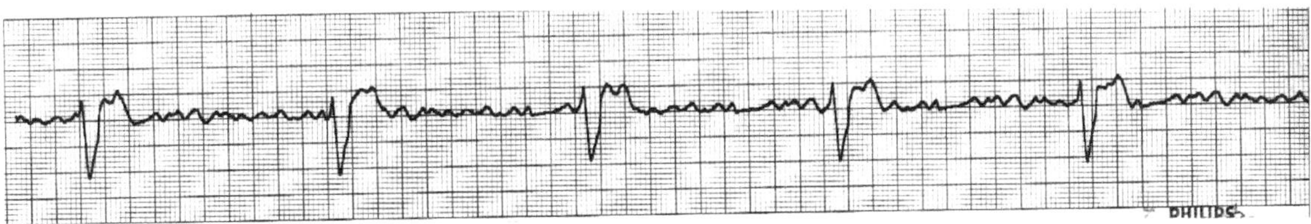

Causes: 3rd degree AVB may be transient and reversible or permanent. Common causes are: acute inferior or right ventricle MI, ischemic heart disease in general, increased vagal tone, digitalis toxicity, amiodarone, beta-blockers, calcium channel blockers and electrolyte imbalances.

Treatment: Signs and symptoms are similar to that of symptomatic sinus bradycardia. This rhythm can progress to ventricular asystole if no back up pacemaker takes over. Pacing is the treatment of choice, at first a temporary pacer may be deployed until the patient can have a permanent pacer implanted.

Idioventricular Rhythm

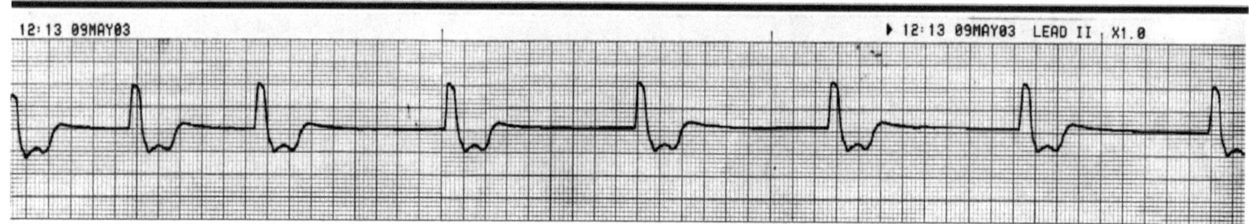

Rate: 20-40 BPM
PRI: P waves may be present if SA node is functional; however, there is no relation to the QRS.
QRS: Wide/Bizarre > 0.12 sec
QT: Rate dependant can be < > 0.40 sec
Rhythm: Regular
Source of pacer: Ventricular/Purkinje Fibers

Characteristics: Cells in the ventricles have an automaticity of a rate of 20-40 beats per minute and can function, as a backup pacemaker should the SA or AV node fail. When the rate of a ventricular rhythm is 20-40, the term idioventricular rhythm is used. The QRS is wider than 0.12 seconds with an inverse relationship between the direction of the QRS and the T wave. This is due to abnormal conduction of the electrical impulse through the ventricles. Normally fast track conduction fibers such as the bundle system and Purkinje fibers are utilized and result in a narrow QRS of 0.12 seconds or less. In rhythms of ventricular origin, conduction occurs in a cell-to-cell format resulting in slower conduction, represented by a wide QRS.

Atrial kick is lost with ventricular rhythms, combined with a slower heart rate, the patient may present with signs and symptoms of decreased cardiac output (hypotension, altered mental status, syncope, chest pain, dizziness). If the ventricular rate is fast enough, the patient may present as stable and asymptomatic. **<u>It is worth noting that an idioventricular rhythm is often the last rhythm before asystole and patient death.</u>**

Causes: Myocardial infarction, ischemia, Digitalis toxicity, pacemaker failure and metabolic imbalances.

Treatment: Finding and treating reversible causes and use of a temporary pacer are the golden standards of treatment. Atropine and isoproterenol may be used to try to increase the heart rate. Medications such as lidocaine or other antiarrhythmics may further decrease ventricular rate and rhythm.

Accelerated Idioventricular Rhythm

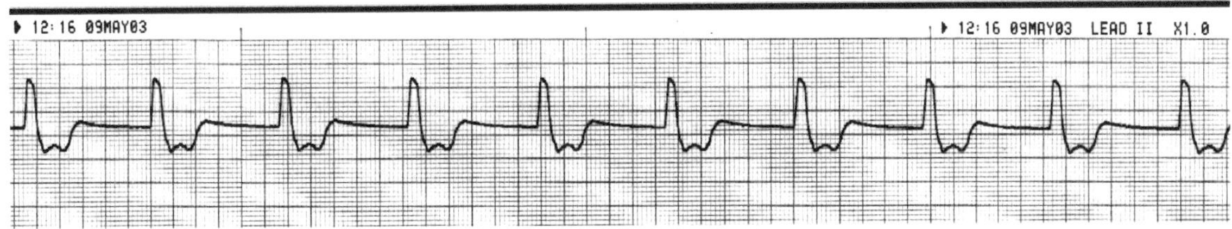

Rate: 40-100 BPM
PRI: P waves may be present if the SA node is functional; however, there is no relation to the QRS.
QRS: Wide/Bizarre > 0.12 sec
QT: Rate dependant can be < > 0.40 sec
Rhythm: Regular
Source of pacer: Ventricular/Purkinje Fibers

Characteristics: Characteristics of an accelerated idioventricular rhythm are similar to that of an idioventricular rhythm. An accelerated idioventricular rhythm will have a rate of 40-100 BPM, with a wide QRS and an inverse relationship between the direction of the QRS and T wave.

Causes: Acute MI, ischemia, digitalis toxicity, metabolic disorders. This rhythm may also occur as a result of thrombolytic therapy.

Treatment: Accelerated Idioventricular rhythms can be self-terminating and harmless and may only require treatment if the patient is symptomatic. However, this rhythm may be a prognostic sign that a more serious condition is present such as: drug toxicity, AMI, ischemia, CAD and electrolyte disorders. This Treatment is similar to that of an idioventricular rhythm. **An idioventricular rhythm is often the last rhythm before asystole and patient death.**

Ventricular Tachycardia

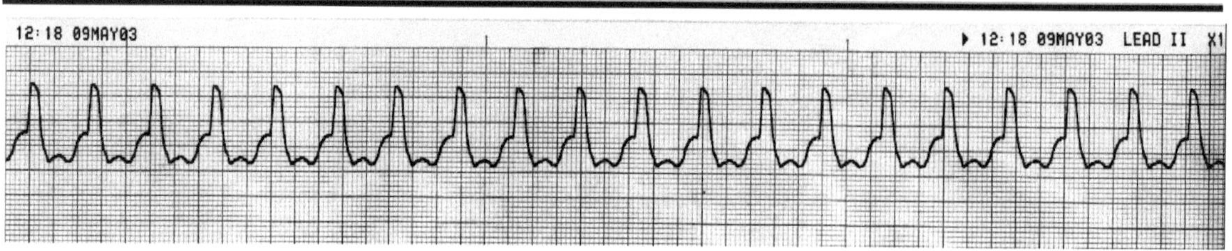

Rate: > 100 BPM
PRI: P waves may be present if SA node is functional; however, there is no relation to the QRS. It is unlikely that P waves will be visible, since this rhythm typically moves at a very fast rate, and the P waves will be buried in the QRS.
QRS: Wide/Bizarre > 0.12 sec
QT: Rate dependant may be < 0.40 seconds with a fast rhythm.
Rhythm: Regular
Source of pacer: Ventricular/Purkinje Fibers

Characteristics: VT is a rhythm of ventricular origin with a rate faster than 100 beats per minute. VT can appear as polymorphic (different QRS complexes) or monomorphic (QRS complexes are all the same). VT results form an ectopic focus or a ventricular reentry pathway.

Depending on the rate of the tachycardia, the patient can present as stable or unstable. If VT occurs at a relatively slow rate in a healthy heart, the rhythm can be well tolerated. However, if the rhythm is fast, ventricular diastolic filling time is limited resulting in poor cardiac output and hemodynamic instability. Patients with poor left ventricular function and ejection fraction do not tolerate any form of tachycardia. The longer that a patient is left in sustained VT, the harder it may be to convert to a regular sinus rhythm.

Causes: Ventricular tachycardia may be caused by: R on T PVC phenomenon, hypoxia, ischemia, AMI, acidosis, cardiomyopathy, mitral valve prolapse, digitalis toxicity, antiarrhythmics, electrolyte imbalances, liquid protein diets, increased intracranial pressure and central nervous system disorders.

Treatment: Stable patients are given medications to attempt to chemically convert them. Unstable patients are to be treated promptly with defibrillation and medications. Please consult your ACLS manual for detailed treatment algorithms.

Premature Ventricular Contractions

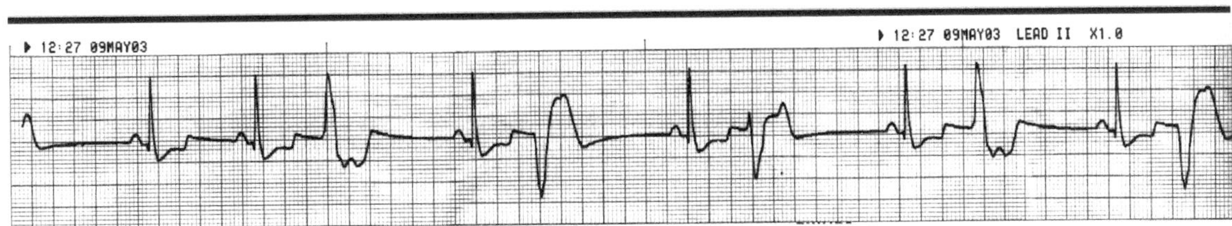

Rate: That of the underlying rhythm
PRI: That of the underlying rhythm
QRS: Will vary depending on the site of pacer of the conducted beat.
QT: Usually, < 0.40 sec
Rhythm: Irregular due to premature beats.
Source of pacer: Ventricular/Purkinje Fibers

Characteristics: PVCs conduct muscle cell to muscle cell in the ventricles, which is much slower than the purkinje system, this results in the wide and bizarre looking QRS, which has an inverse relationship in the direction of the T wave. Retrograde conduction from the PVC back through the AV node and into the atria can occur, resulting in inverted P waves see following the PVC.

PVCs can appear as single, or in couplets or triplets. 6 or more PVCs occurring in a row is considered a run of VT. PVCs may also appear in different shapes indicating multiple sites of ectopy in the ventricles. Alternatively, PVCs may also appear in the same shape as they originate from a single source of ventricular ectopy. The strip above, displays both unifocal and multifocal PVCs.

R on T Phenomenon: This occurs when a PVC fires during the relative refractory period of the T wave (down slope of the T wave). During this period, the myocardium is susceptible to stimulation from a strong electrical impulse. At this time, some of the myocardium is in a transition state of repolarization, some cells are repolarized, some only partially repolarized and others are still depolarized. An electrical stimulus, such as a PVC may throw the heart into electrical chaos causing a lethal arrhythmia such as VT, VF or torsades de pointes. A long QT (> 0.40 sec) in the presence of PVCs places the patient at risk, for the occurrence of an R on T phenomenon.

Causes: Emotions (increased catecholamine), stimulants such as coffee, nicotine, ethanol (ETOH), cocaine, amphetamines, AMI, CHF, digitalis, increased vagal tones, hypoxia, acidosis, hypokalemia, hypomagnesemia, acidosis, ischemia, hypoxia and open-heart surgery.

Treatment: May be treated with beta-blockers, procainamide, lidocaine, amiodarone or other medications. Attention should also be placed on finding and treating underlying causes.

Ventricular Fibrillation

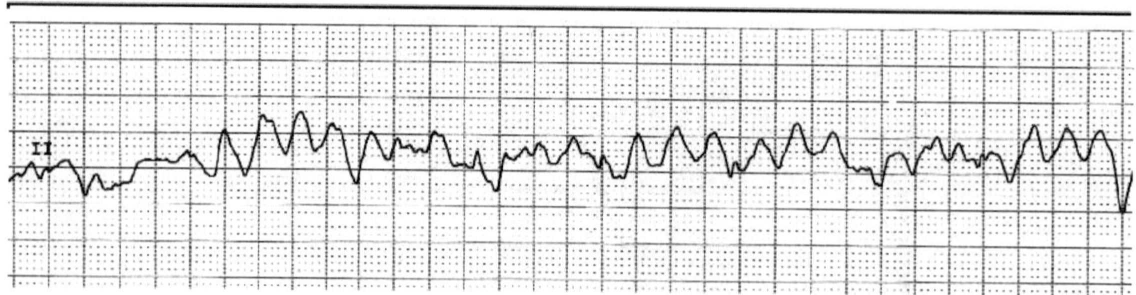

Rate: Unable to measure due to rapid and uncoordinated electrical activity.
PRI: 0
QRS: 0
QT: 0
Rhythm: Rapid and chaotic.
Source of pacer: Multiple sources of ectopy in the Ventricular/Purkinje Fibers

Characteristics: VF is chaotic electrical activity in the ventricles that results in quivering of the ventricles and total loss of cardiac output. VF is fatal unless treated promptly with CPR and defibrillation. There are no QRS complexes or P waves present. VF can be course or very fine and look almost like asystole.

Causes: CAD, AMI, trauma, hypoxia, acidosis, antiarrhythmics, electrolyte imbalances, cardiac catheterization, cardioversion, accidental electrocution, cardiac pacing and extreme hypothermia.

Treatment: CPR, immediate treatment with antiarrhythmic medications and defibrillation with appropriate joule settings. Refer to your ACLS manual for detailed treatment algorithms.

Asystole

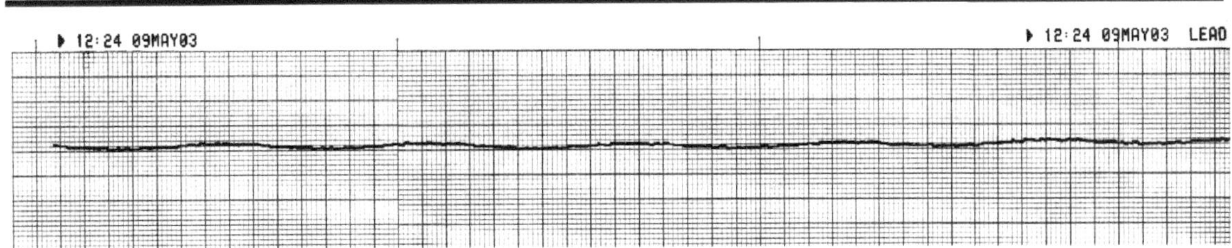

Rate: 0
PRI: 0
QRS: 0
QT: 0
Rhythm: None
Source of pacer: None

Characteristics: Asystole is the sudden loss of ventricular electrical activity, resulting in no ventricular contractions and no cardiac output. Asystole is fatal unless reversed immediately. Asystole should always be confirmed in another monitor lead before treatment. A true asystole will not respond to defibrillation.

Causes: End stage cardiac disease, ischemia, MI, severe electrolyte imbalances, acidosis, and hypoxia.

Treatment: CPR, epinephrine, atropine and external pacing. Refer to ACLS manual for detailed treatment algorithms.

Torsade de Pointes

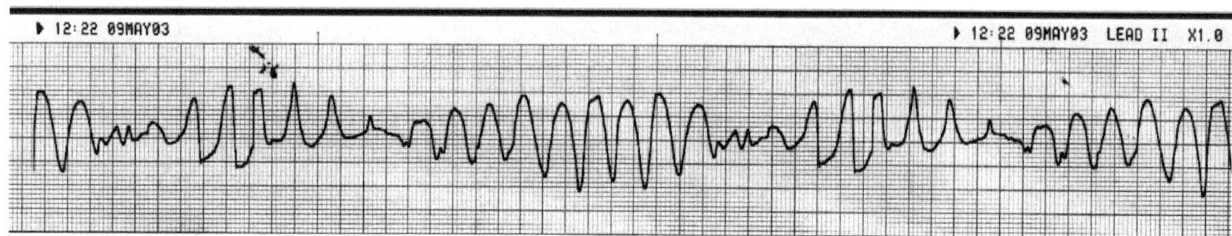

Rate: 100-250 BPM
PRI: P waves may be present if SA node is functional; however, there is no relation to the QRS. It is unlikely that P waves will be visible, since this rhythm typically moves at a very fast rate, and the P waves will be buried in the QRS.
QRS: Wide/Bizarre > 0.12 sec
QT: Usually, < 0.40 seconds due to fast heart rate.
Rhythm: Usually regular
Source of pacer: Ventricular/Purkinje Fibers

Characteristics: Torsade de pointes is essentially a polymorphic VT. It is characterized by a widening and narrowing of the QRS amplitude. The arrhythmia may be paroxysmal, which starts and stops suddenly and may suddenly deteriorate into VF.

Causes: The cause of torsades may be reversible. The most common causes are drugs that lengthen the QT interval such as antiarrhythmics (quinidine, procainamide and sotalol for example). Other causes include myocardial ischemia, and hypokalemia, hypomagnesemia and hypocalcemia.

Treatment: Find and treat reversible causes. Overdrive pacing with the use of an external pacer or isoproterenol may overdrive the ventricular rate and break the triggering mechanism of the arrhythmia. Magnesium sulfate may also be effective. Refer to your ACLS manual for detailed treatment algorithms.

Pacemaker Rhythms

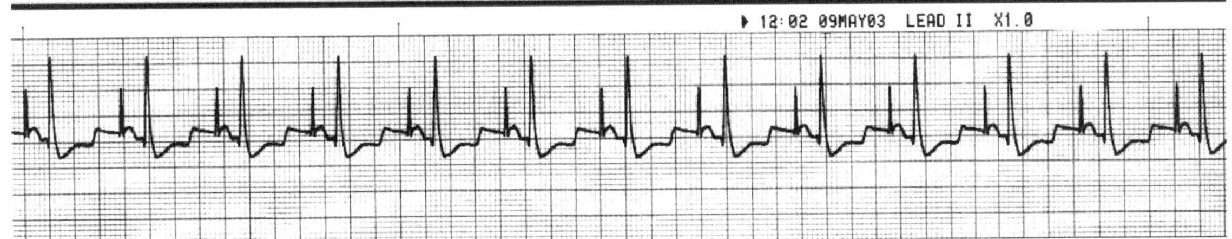

Single chamber atrial pacer with 100% capture.

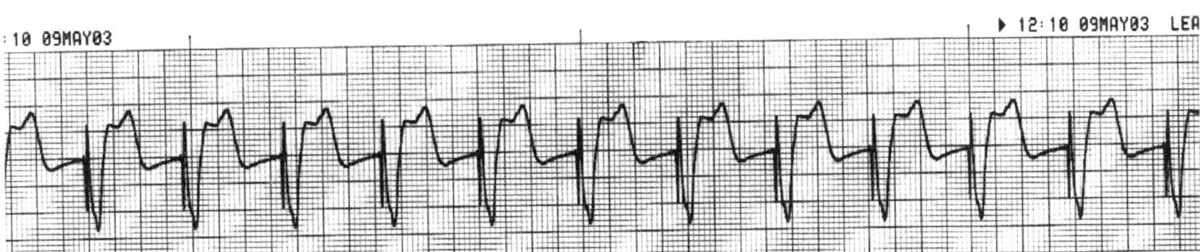

Single chamber ventricular pacer with 100% capture.

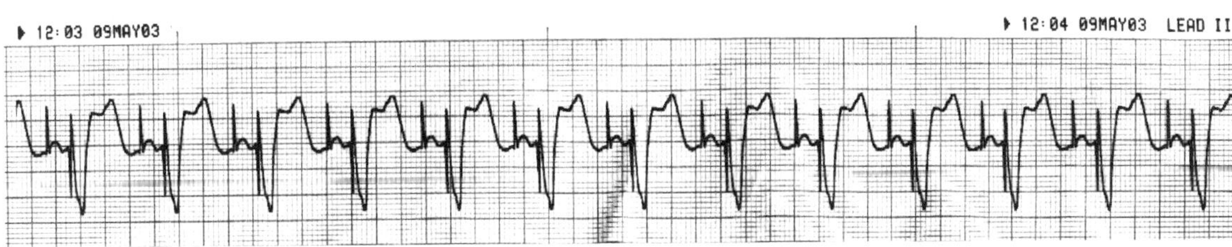

Dual chambered (A/V sequential) pacer with 100% capture.

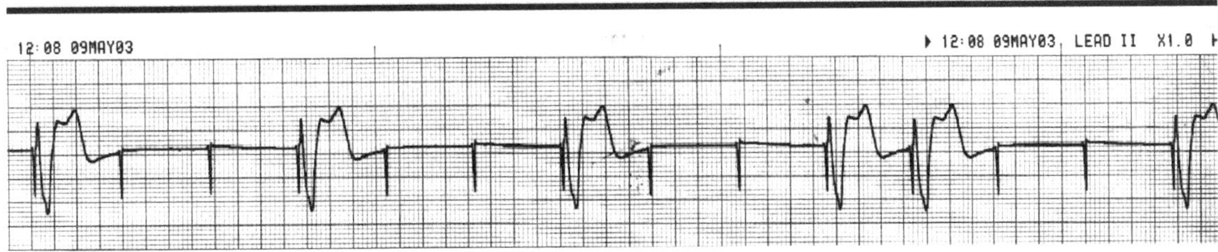

Single chamber ventricle pacer with intermittent loss of capture.

Rate: Dependent on rate of pacer and spontaneous rhythm.
PRI: When paced, dependent on type of pacer and settings. May be variable.
QRS: Dependent on pacer type and site.
QT: Variable
Rhythm: Typically regular, but may be irregular with a malfunctioning pacer.
Source of pacer: May be intrinsic or form the pacemaker.

Characteristics: Permanent implanted pacemakers come in a variety of types such as single or dual chambered pacers, and operate in different modes such as demand or fixed.

Atrial chambered pacers: The SA node typically is damaged and nonfunctional. A pacer lead is placed in the atria to produce an electrical stimulus to drive the atria. The atrial pacer impulse is conducted through the AV node, bundle branches and purkinje fibers in a normal fashion. Each P wave is preceded by a pacer spike. Atrial kick is maintained, since the pacer causes atrial contraction and allows the atria to contribute to cardiac output.

AV sequential pacers: These are dual chambered pacers that drive both the atria and ventricles. Each P wave is preceded by a pacer spike, and each QRS is preceded by a pacer spike. Atrial kick is maintained.

Ventricular pacers: are typically older style pacers in which a lead is placed in the ventricles and serves as the electrical impulse generator that causes the ventricles to contract. Since only the ventricles are contracting, atrial kick is lost and cardiac output is reduced by an estimated 20-30%.

When evaluating a paced rhythm, each pacer spike should have a response from either the atria or ventricles in the form of a P wave or QRS. If there are more pacer spikes than appropriate waveforms, it is said that capture is being lost. If there is a waveform for each and every pacer spike, the pacer is functioning correctly with 100% capture.

Failure to pace can be seen on the EKG strip as a pause, in which case a pacer spike should have occurred. This is caused by pacer circuit or battery failure and can lead to asystole.

Failure to sense is seen when a demand mode pacer fails to sense the patient's intrinsic rate and fires a pacer spike on top of a QRS or other waveform. This can be viewed as the pacer trying to help when no help is needed.

Implanted permanent pacers can be interrogated and adjusted externally through the use specialized equipment.

Temporary external pacers can have the rate, mode and sensitivity easily adjusted to achieve 100% capture and optimal sensing.

Treatment: Malfunctioning implanted pacers cannot be fixed at the bedside without the use of specialized equipment. Often the patient is taken to the cardiac catheterization lab to have the pacer interrogated and replaced as necessary. It is important to contact the physician and report any changes in the patient's paced rhythm or change in physical condition.

Summary of Heart Rhythms

Atrial Rhythms
Premature Atrial Contraction
Wandering Atrial Pacemaker
Atrial Tachycardia
Atrial Flutter
Atrial Fibrillation
Multifocal Tachycardia
Supraventricular Tachycardia

Sinus Rhythm
Normal Sinus Rhythm
Sinus Tachycardia
Sinus Bradycardia
Sinus Arrhythmia
Sinus Pause
Sinus Arrest

Junctional Rhythms
Premature Junctional Contraction
Junctional Rhythm
Accelerate Junctional Rhythm
Junctional Tachycardia

Atrial Ventricular Blocks
First Degree
Second Degree Type I
Second Degree Type II
Third Degree Heart Block

Ventricular Rhythms
Premature Ventricular Contraction
Indioventricular Rhythm
Accelerated Ventricular Rhythm
Ventricular Tachycardia
Ventricular Fibrillation
Torsade de Points

References

Aehlert, A. (2002). *ECG's made easy* (2nd ed.). St. Louis: Mosby Year Book Inc..

Ambrose, M., Mauro, E., Nash, J., & Sinovic, D. (1997). *ECG Interpretation made incredibly easy.* Springhouse, PA: Springhouse.

American Association of Critical Care Nurses (Ed.). (2001). *Basic EKG Module* (2nd ed., Vol. 1). Aliso Viejo, CA: American Association of Critical Care Nurses.

Conover, M. (1992). *Understanding electrocardiography arrhythmias and the 12 lead EKG* (6th ed.). St. Louis: Mosby Year Book Inc..

Dubin , D. (1998). *Rapid interpretation of EKG's* (5th ed.). Tampa: COVER Publishing Company.

Grauer, K., & Cavallaro, D. (1997). *Arrhythmia interpretation: ACLS preparation and clinical approach.* St. Louis: Mosby Lifeline.

Huff, J. (1997). *ECG workout: Exercises in arrhythmia interpretation* (3rd ed.). Philadelphia: Lippincott.

Huszar, R. (1994). *Basic dysrhythmia interpretation and management.* St. Louis: Mosby Year Bokk Inc..

Huszar, R. (2002). *Basic dysrhythmias: Interpretation and management* (3rd ed.). St. Louis: Mosby.

Jones, K., & Ochs, G. (1990). *Interpretation of the electrocardiogram: A review for health professionals* (2nd ed.). Norwalk, CT: Appleton & Lange.

Posey, A. (2002). *Basic EKG dysrhythmia identification.* Retrieved November 18, 2004, from http://www.rnceus.com/course_frame.asp?exam_id=16&directory=ekg

Notes

Notes

Basic EKG Work Book

Instructions:

For each EKG strip, measure the PRI, QRS, and QT Segment and identify the rhythm. Please indicate if the ST segment is elevated, elevated, depressed or normal. Label the structures of the heart, the cardiac conduction system and coronary arteries.

Cardiac Structures

Label all parts of the heart.

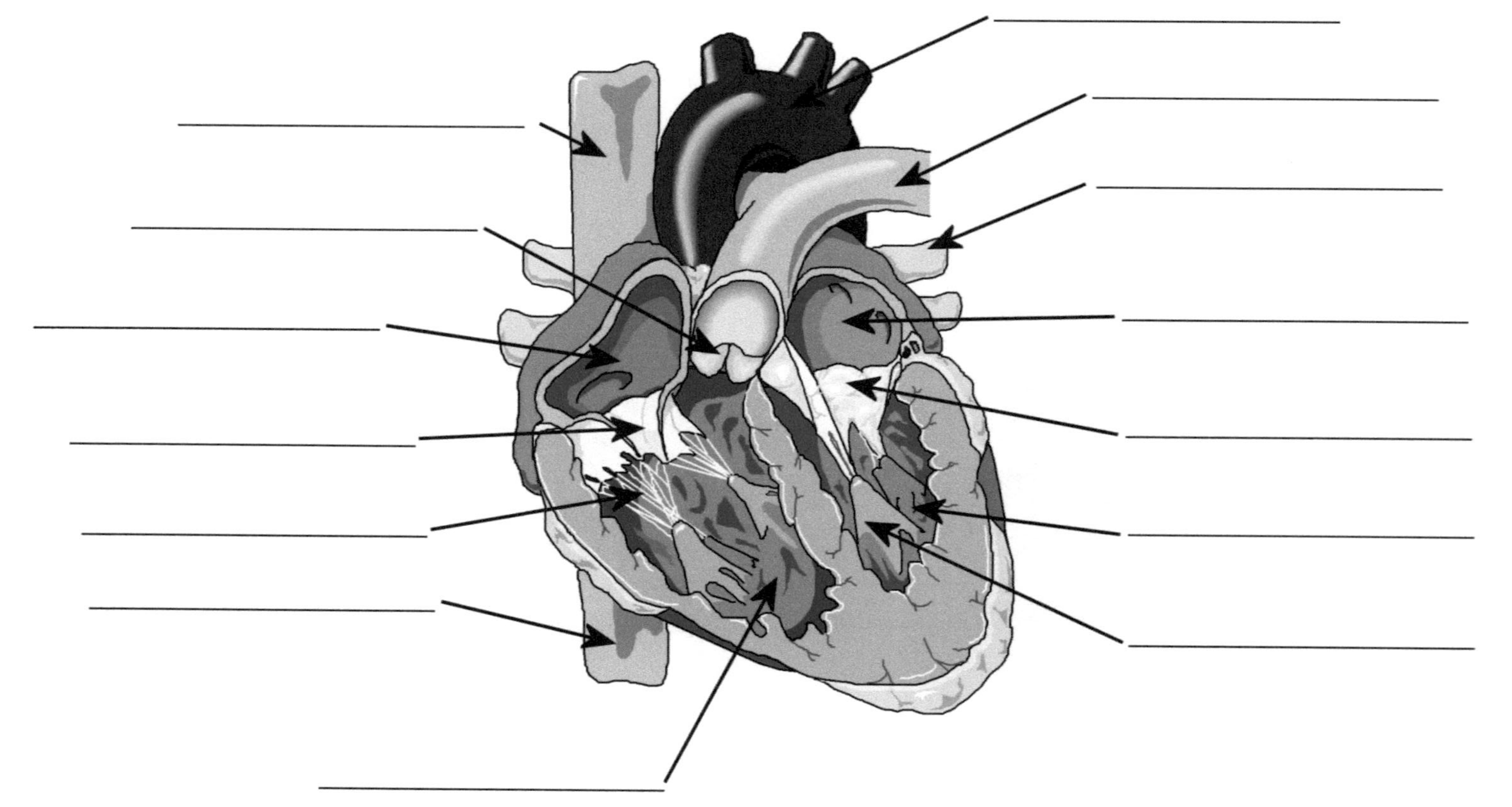

Cardiac Conduction System

Label the components of the cardiac conduction system.

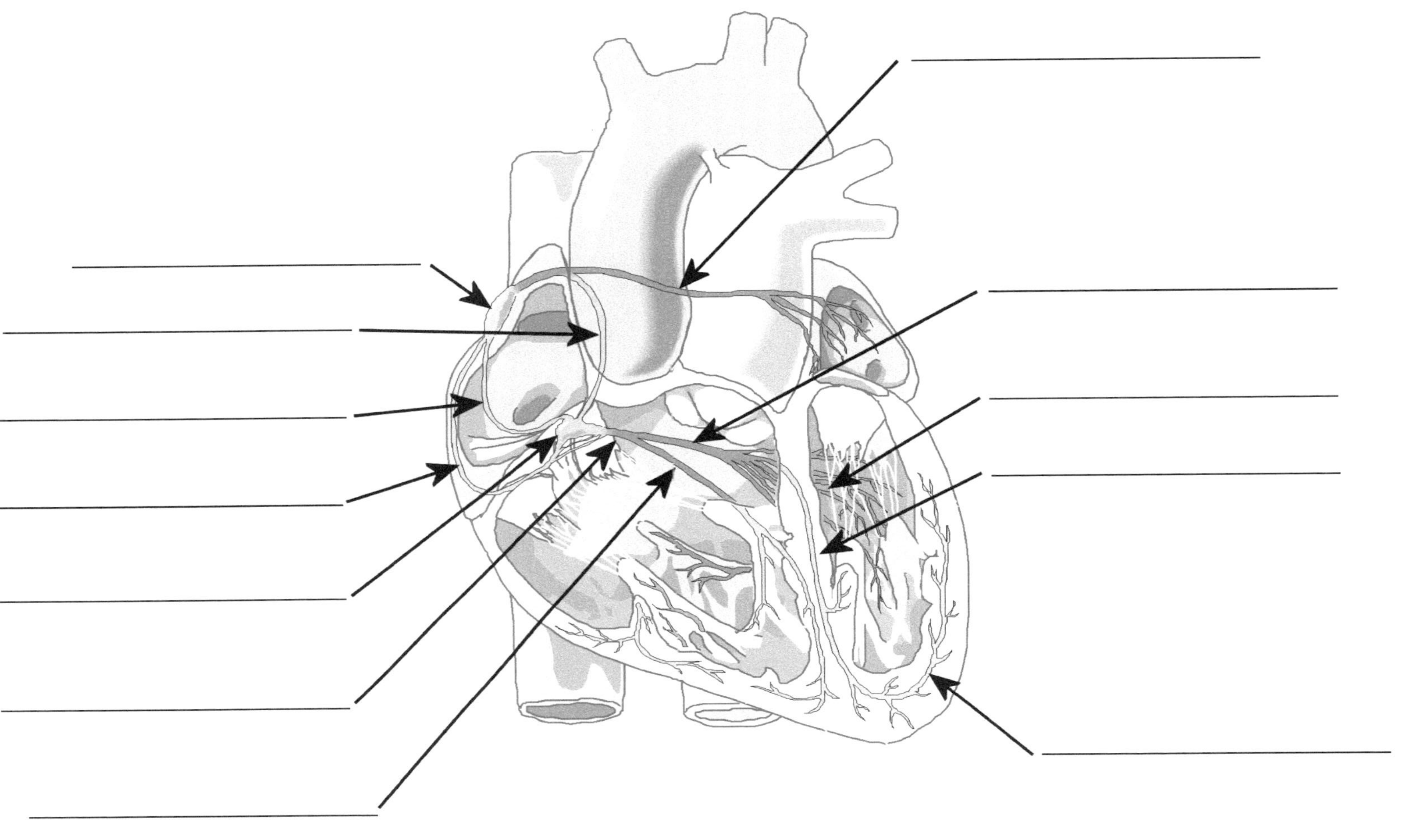

Coronary Arteries:

Fill in the blanks and label all coronary arteries.

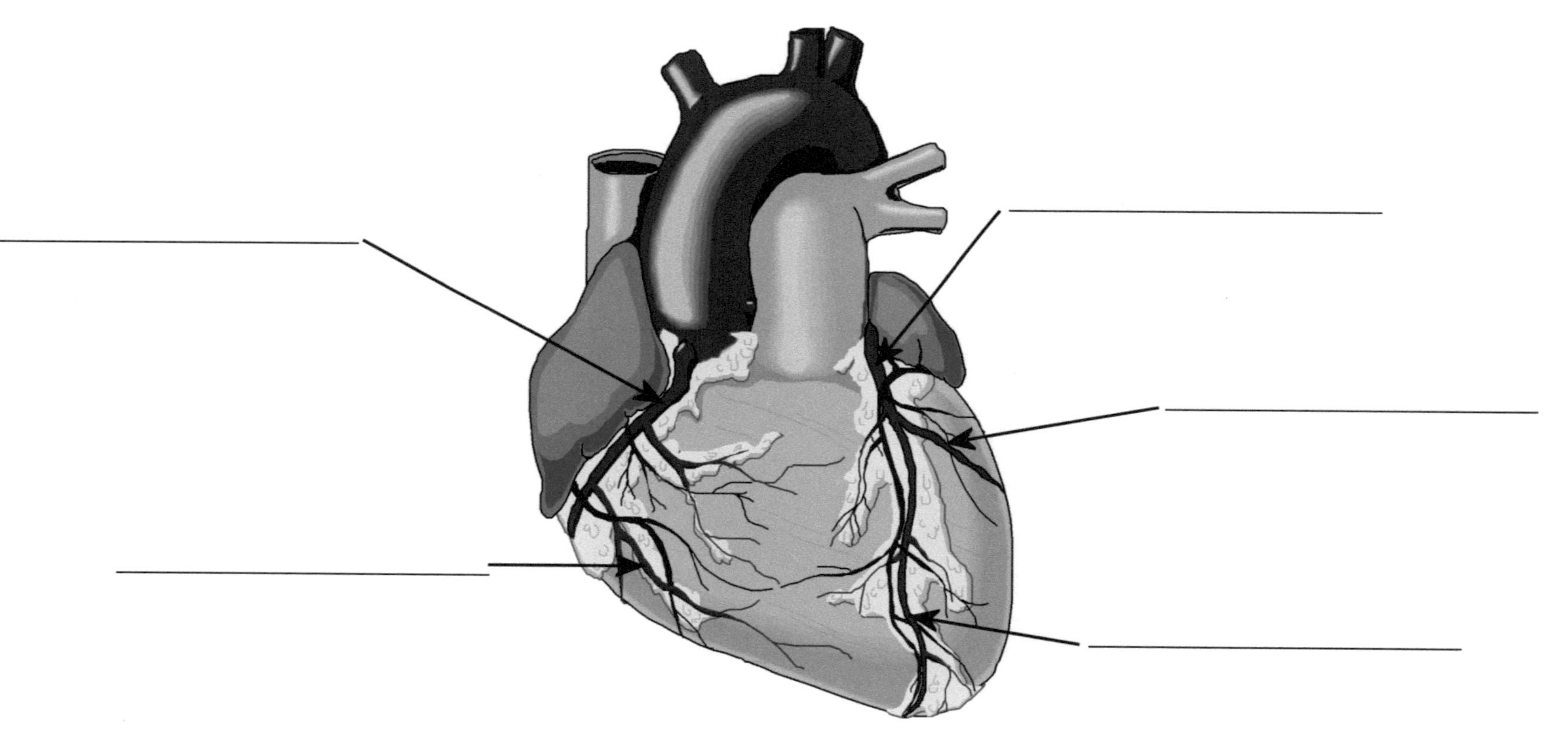

SINUS

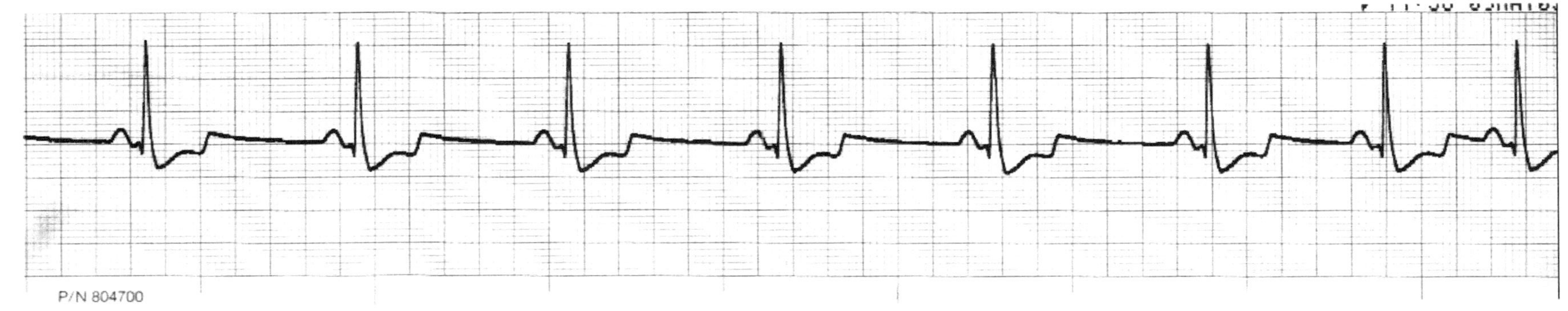

1. PRI:________ QRS:________ QT:_________ Rate:________
   Rhythm:____

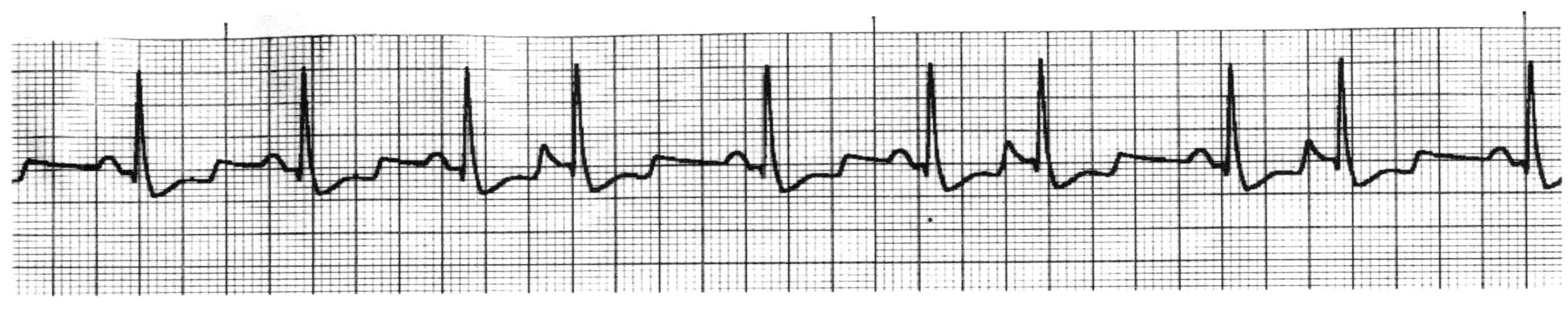

2. PRI:________ QRS:________ QT:_________ Rate:________
   Rhythm:____

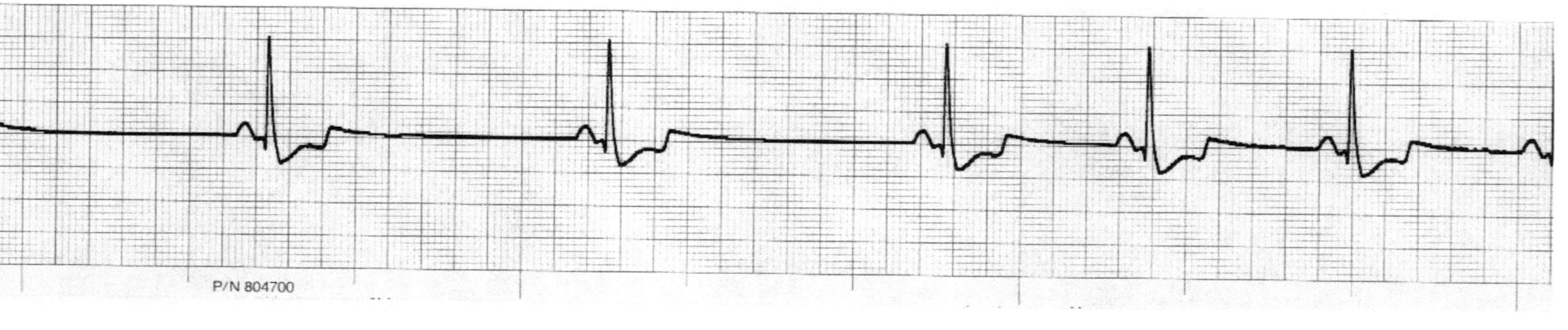

3. **PRI:_________ QRS:_________ QT:__________ Rate:_________**
Rhythm:____

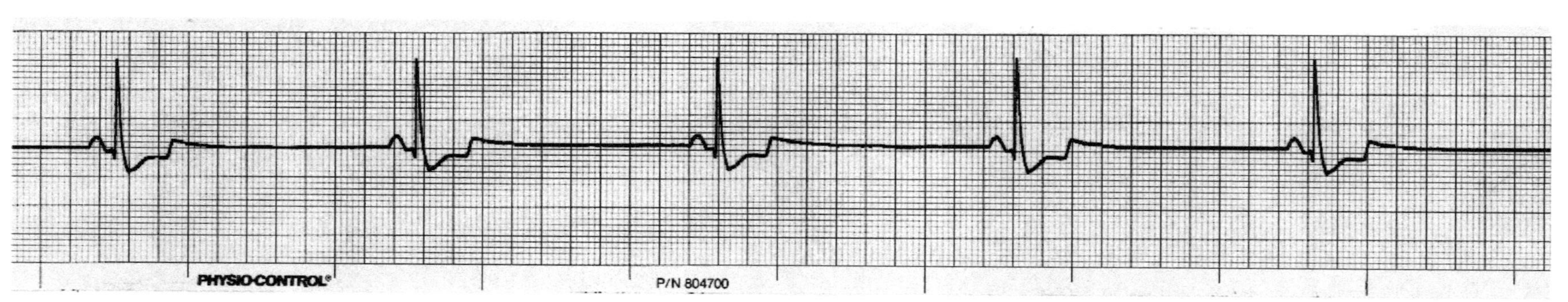

4. **PRI:_________ QRS:_________ QT:__________ Rate:_________**
Rhythm:____

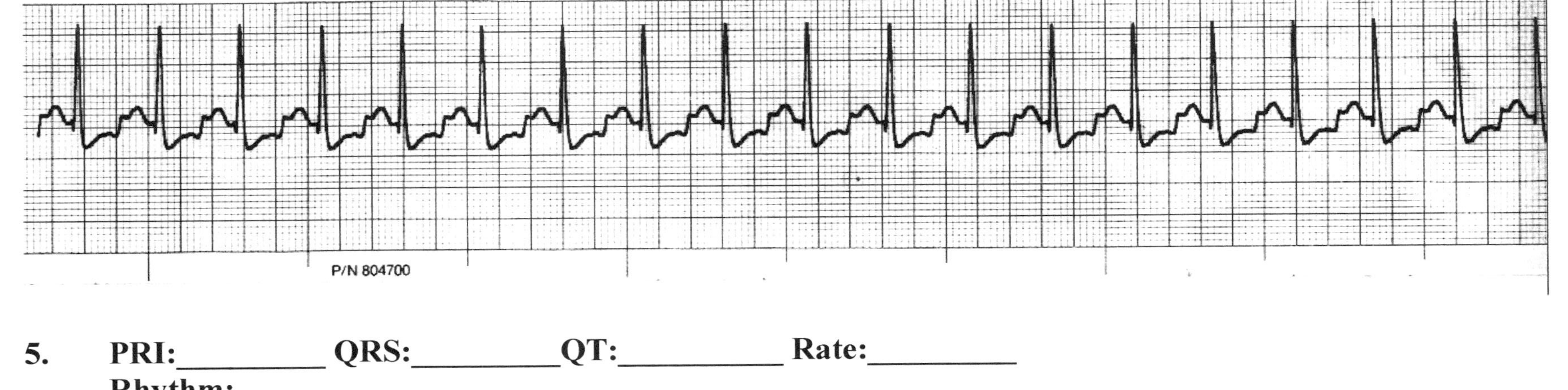

5. **PRI:_________ QRS:_________ QT:_________ Rate:_________**
Rhythm:____

<u>Notes:</u>

Junctional

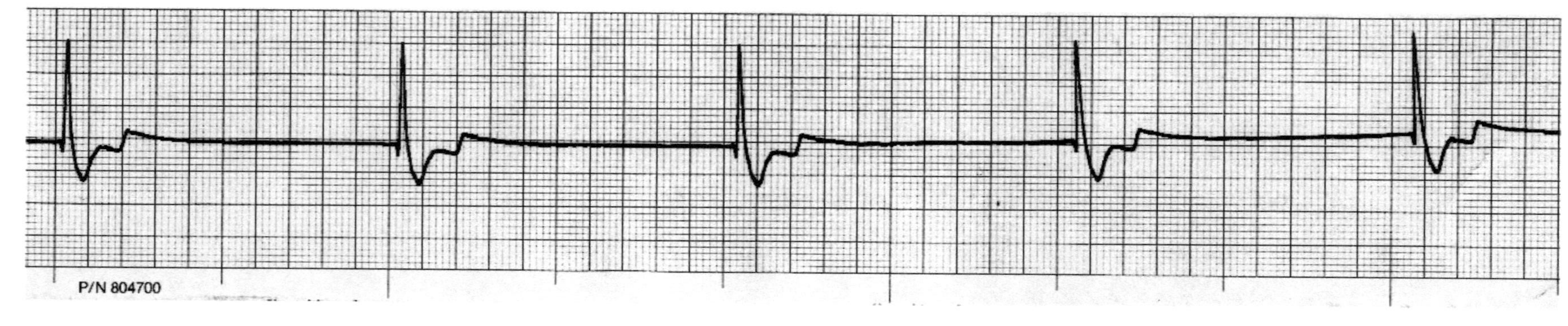

6. PRI:_________ QRS:_________ QT:__________ Rate:_________
Rhythm:____

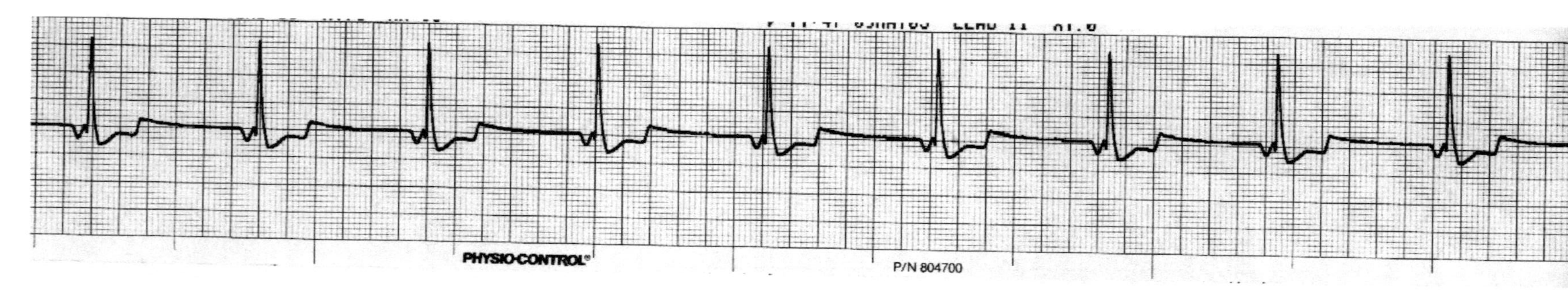

7. PRI:_________ QRS:_________ QT:__________ Rate:_________
Rhythm:____

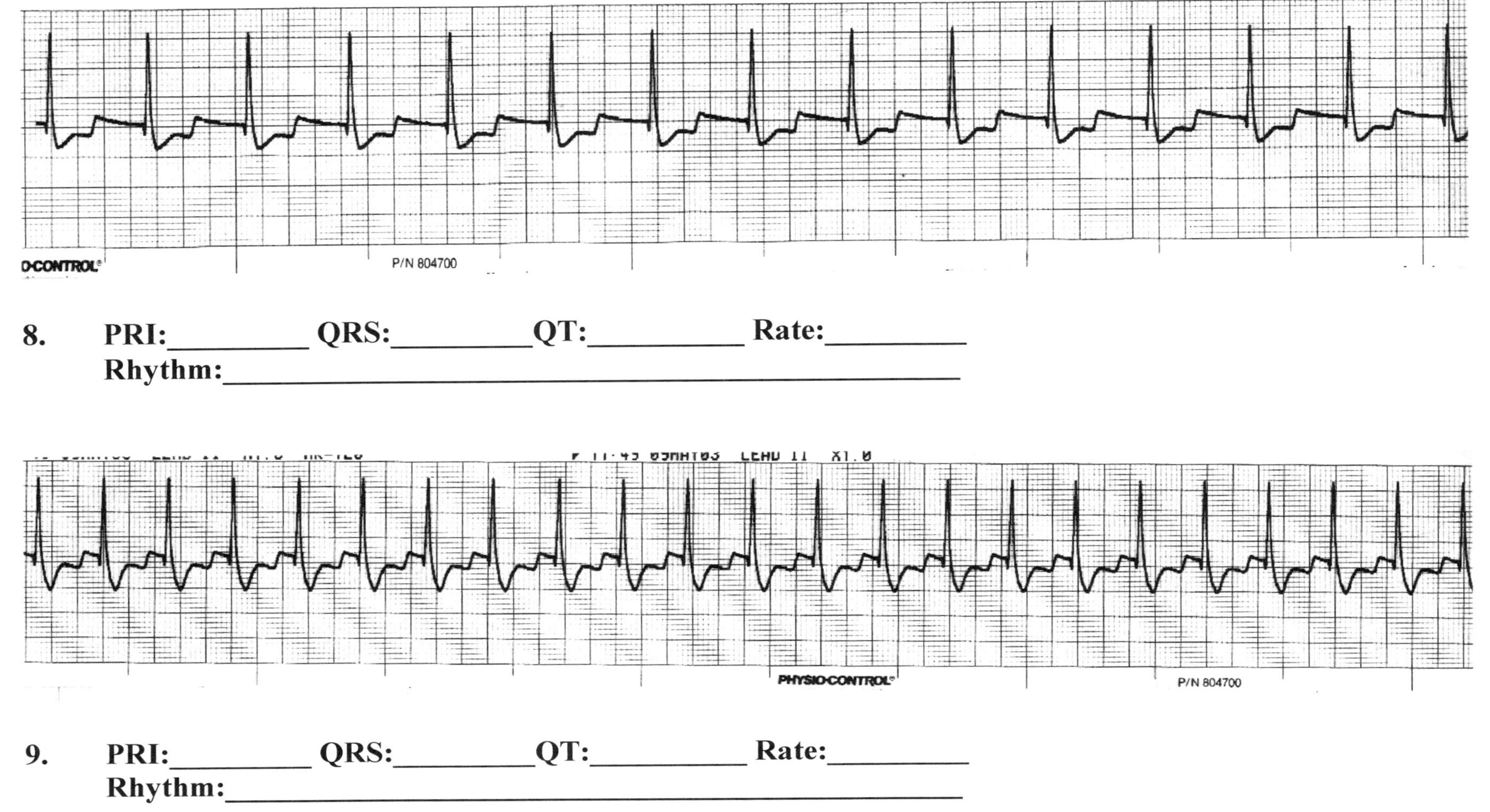

8. **PRI:**_________ **QRS:**_________ **QT:**__________ **Rate:**_________

Rhythm:____

9. **PRI:**_________ **QRS:**_________ **QT:**__________ **Rate:**_________

Rhythm:____

Atrial

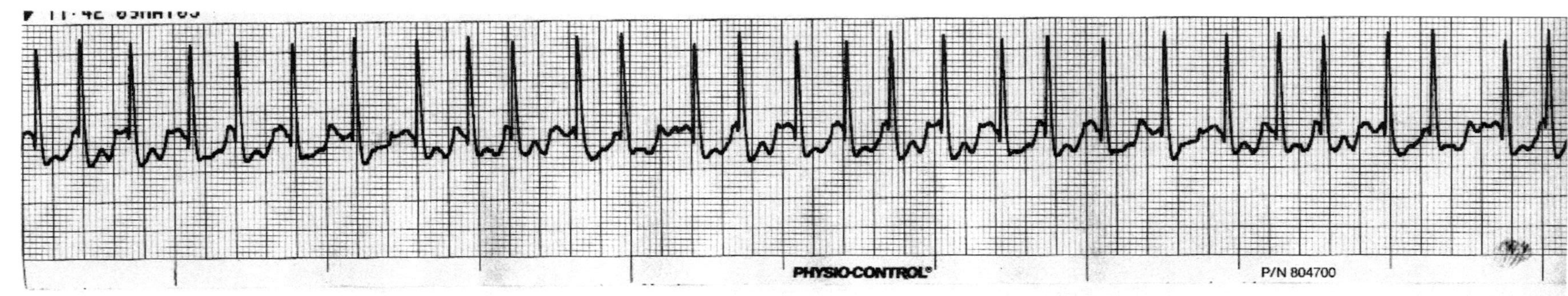

10. PRI:_________ QRS:_________ QT:__________ Rate:_________
Rhythm:____

11. PRI:_________ QRS:_________ QT:__________ Rate:_________
Rhythm:____

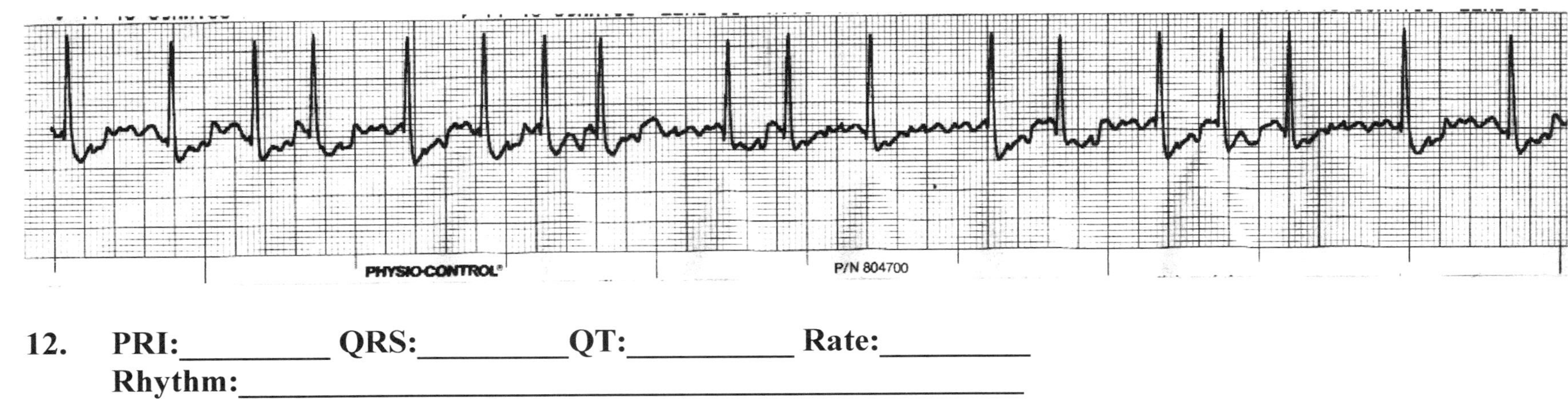

12. PRI:________ QRS:________ QT:_________ Rate:________
Rhythm:____

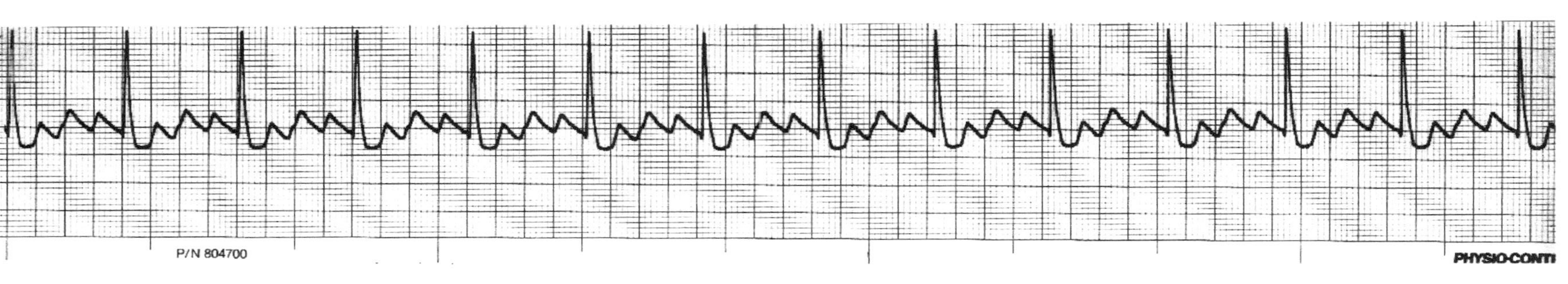

13. PRI:________ QRS:________ QT:_________ Rate:________
Rhythm:____

Ventricular

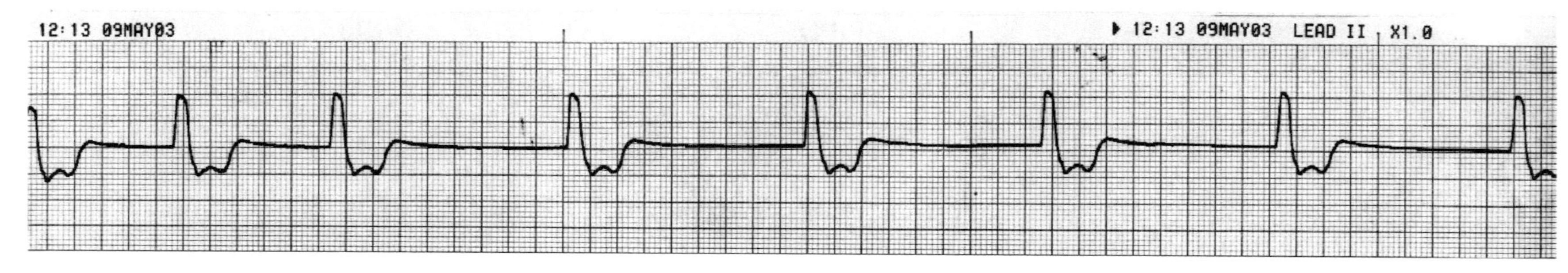

14. PRI:_________ QRS:_________ QT:_________ Rate:_________
Rhythm:____

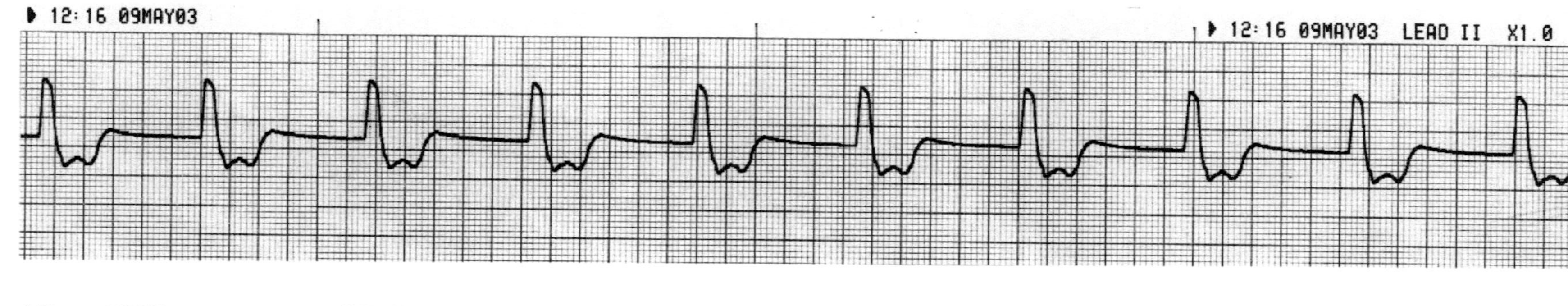

15. PRI:_________ QRS:_________ QT:_________ Rate:_________
Rhythm:____

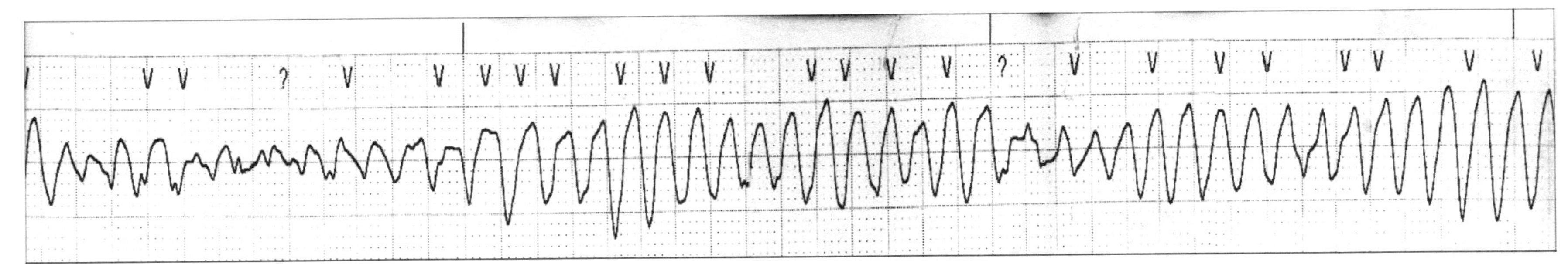

16. PRI:_________ QRS:_________ QT:__________ Rate:_________
Rhythm:____

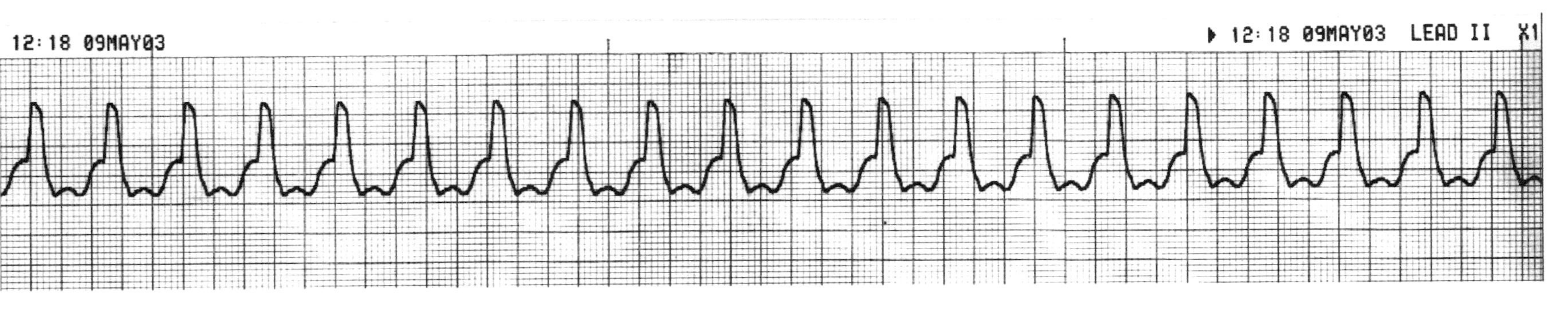

17. PRI:_________ QRS:_________ QT:__________ Rate:_________
Rhythm:____

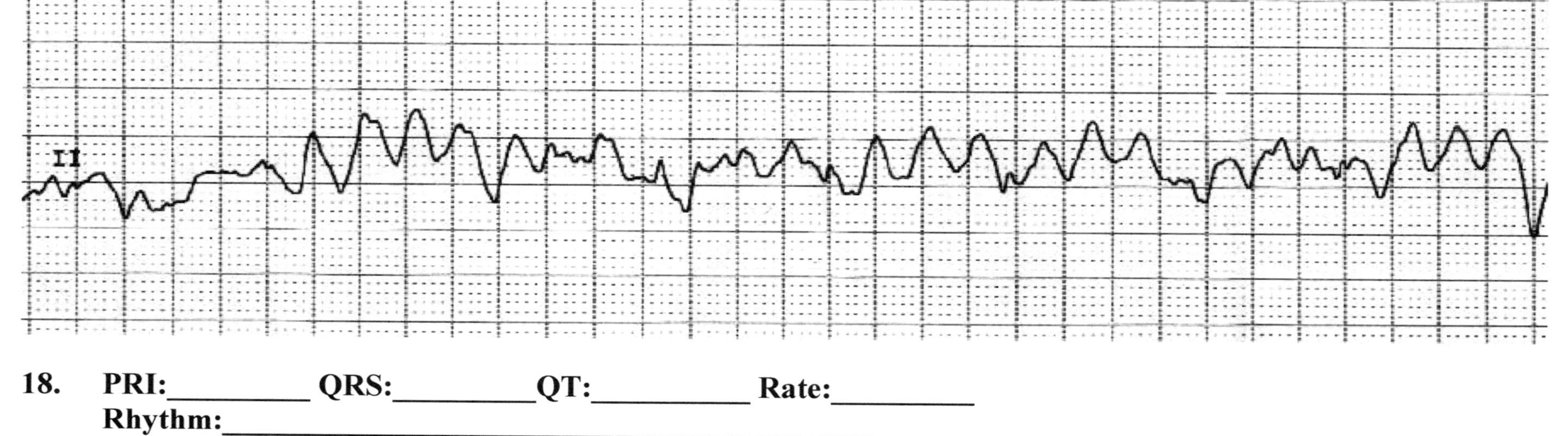

18. PRI:_________ QRS:_________ QT:_________ Rate:_________
Rhythm:____

Notes:

AV Blocks

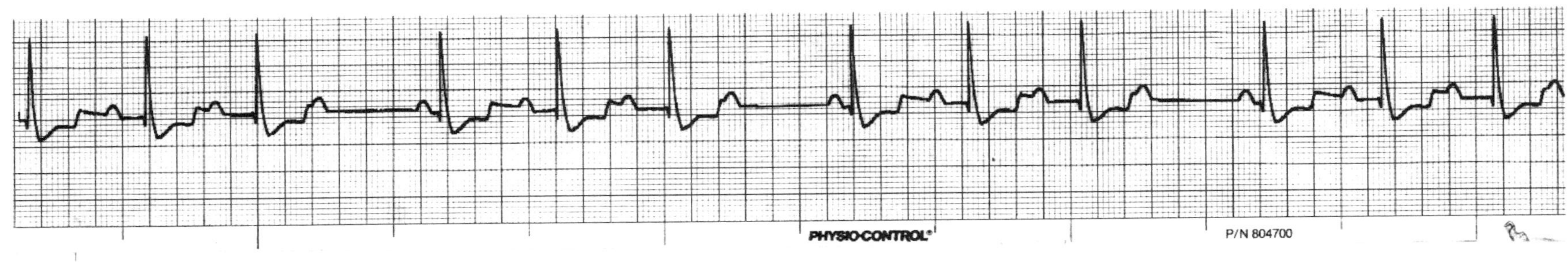

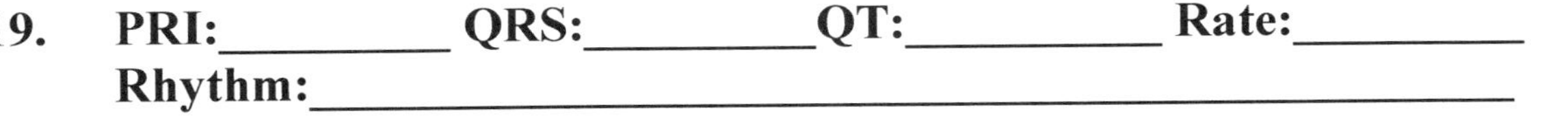

19. **PRI:_________ QRS:_________ QT:__________ Rate:_________**
Rhythm:____

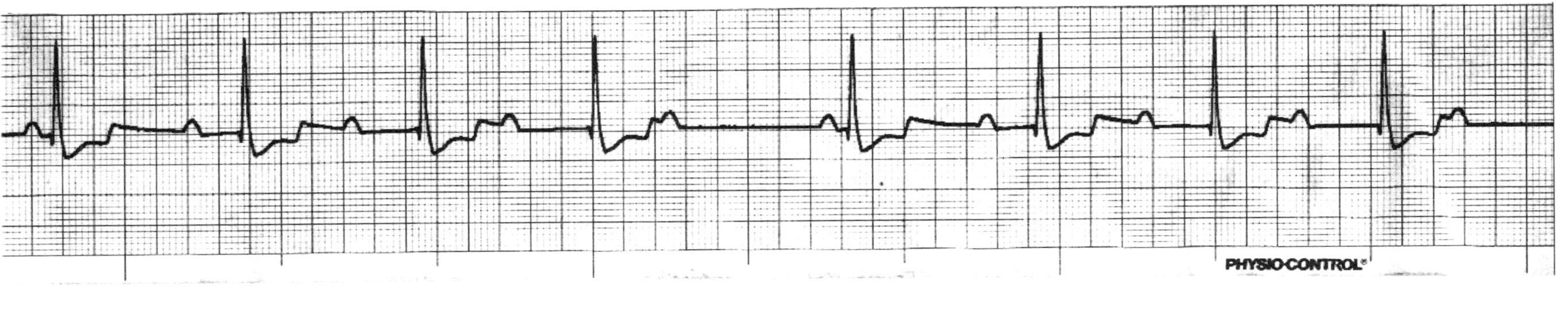

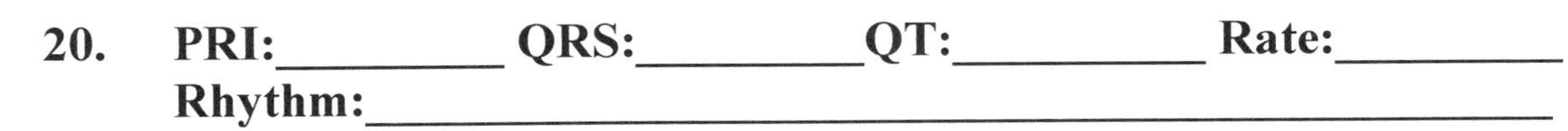

20. **PRI:_________ QRS:_________ QT:__________ Rate:_________**
Rhythm:____

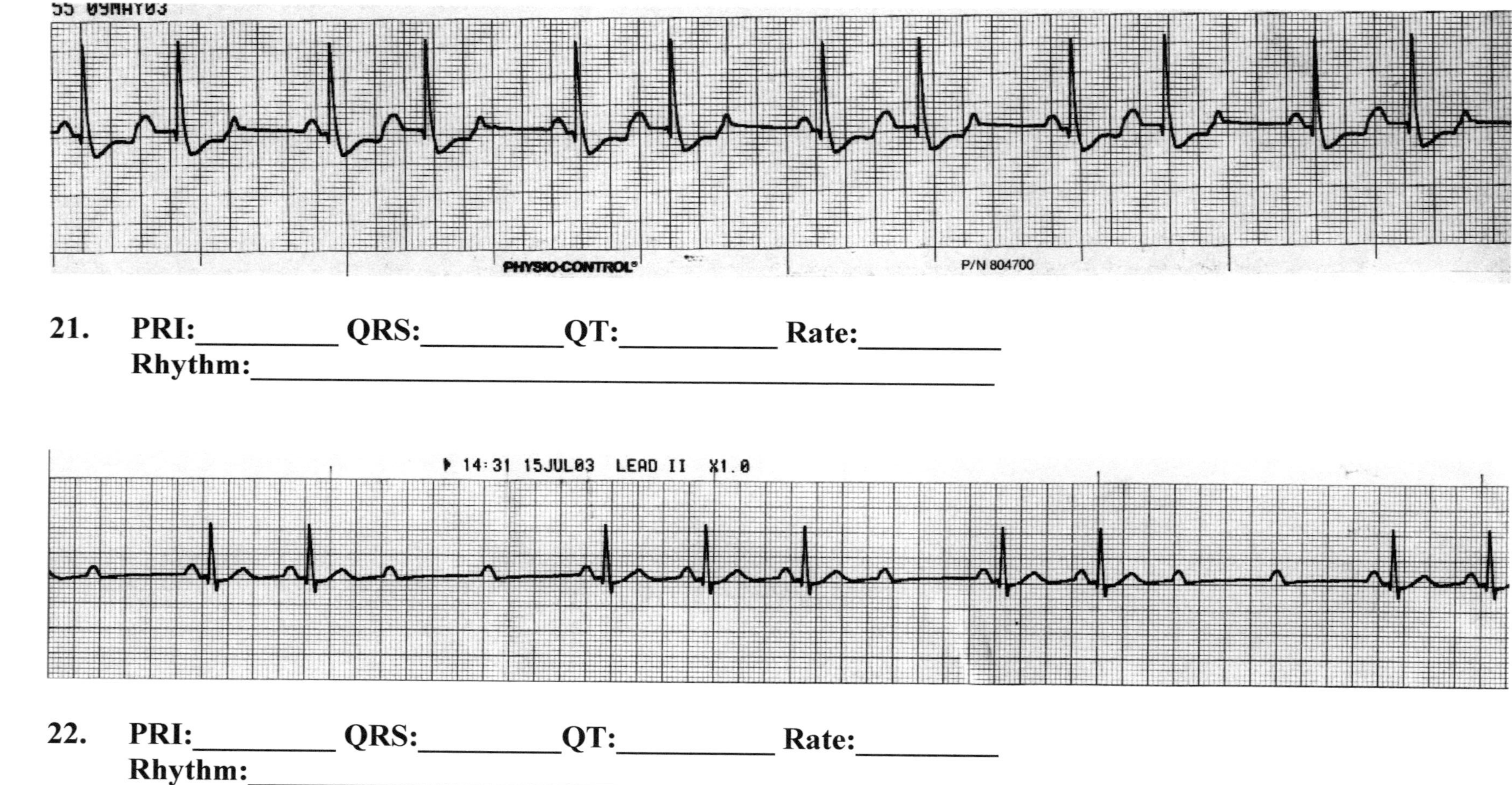

21. PRI:________ QRS:________ QT:_________ Rate:________
Rhythm:____

22. PRI:________ QRS:________ QT:_________ Rate:________
Rhythm:____

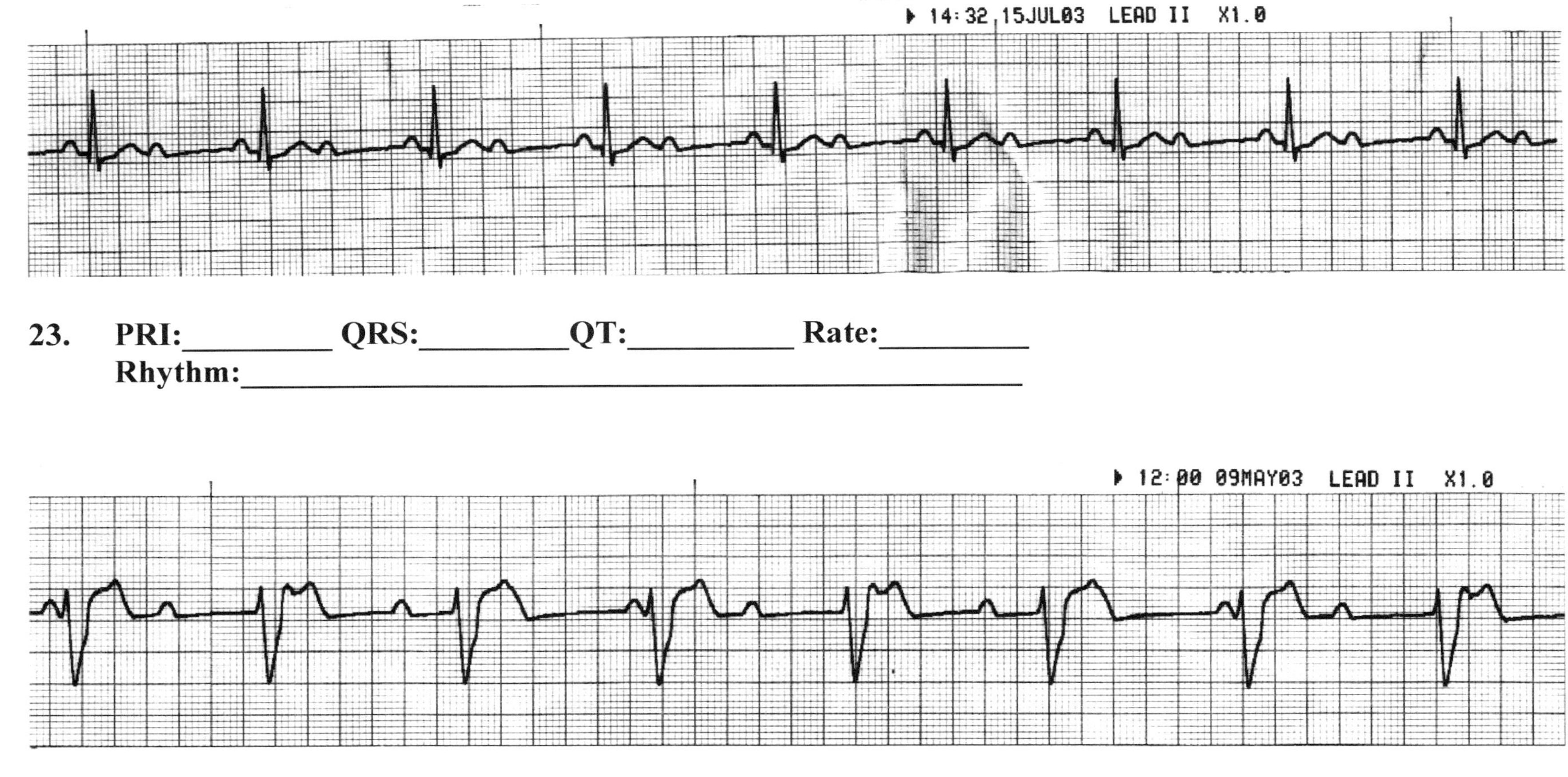

23. **PRI:**________ **QRS:**________ **QT:**________ **Rate:**________
Rhythm:________________________________

24. **PRI:**________ **QRS:**________ **QT:**________ **Rate:**________
Rhythm:________________________________

Pacers

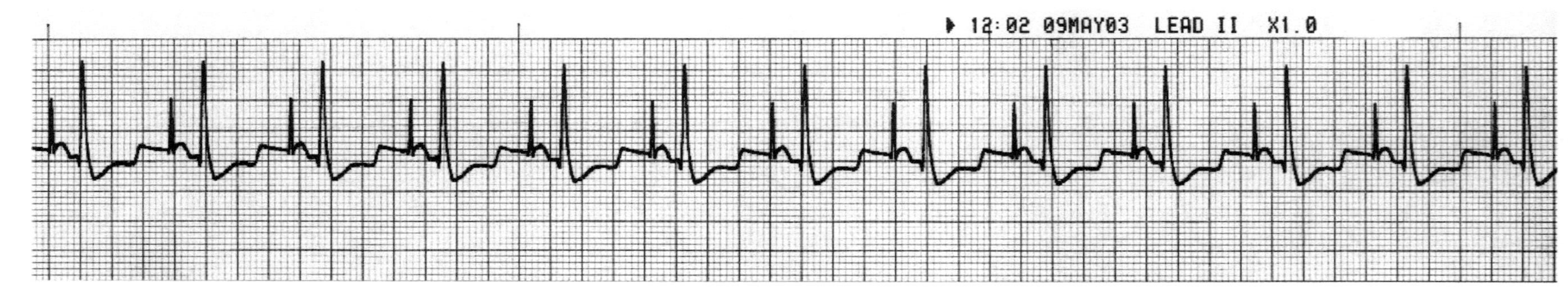

25. PRI:_________ QRS:_________ QT:__________ Rate:_________
Rhythm:____

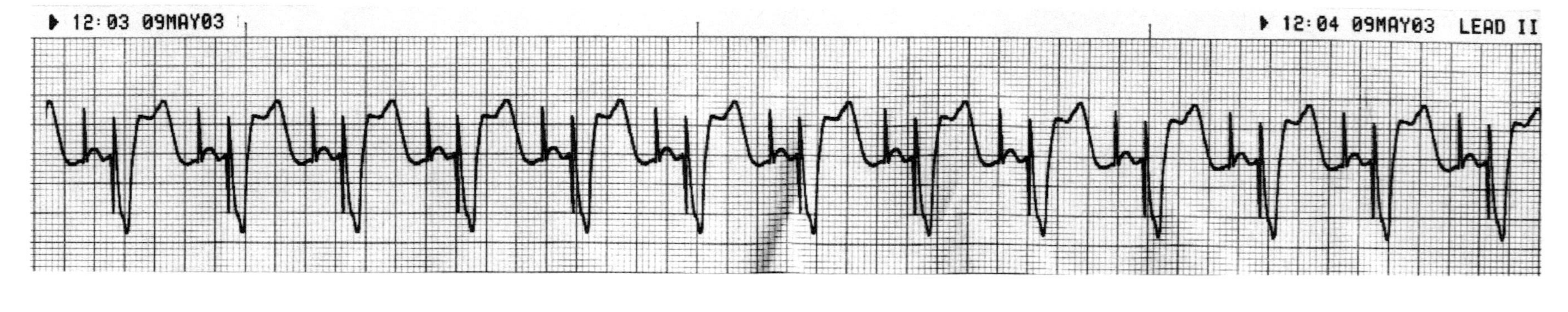

26. PRI:_________ QRS:_________ QT:__________ Rate:_________
Rhythm:____

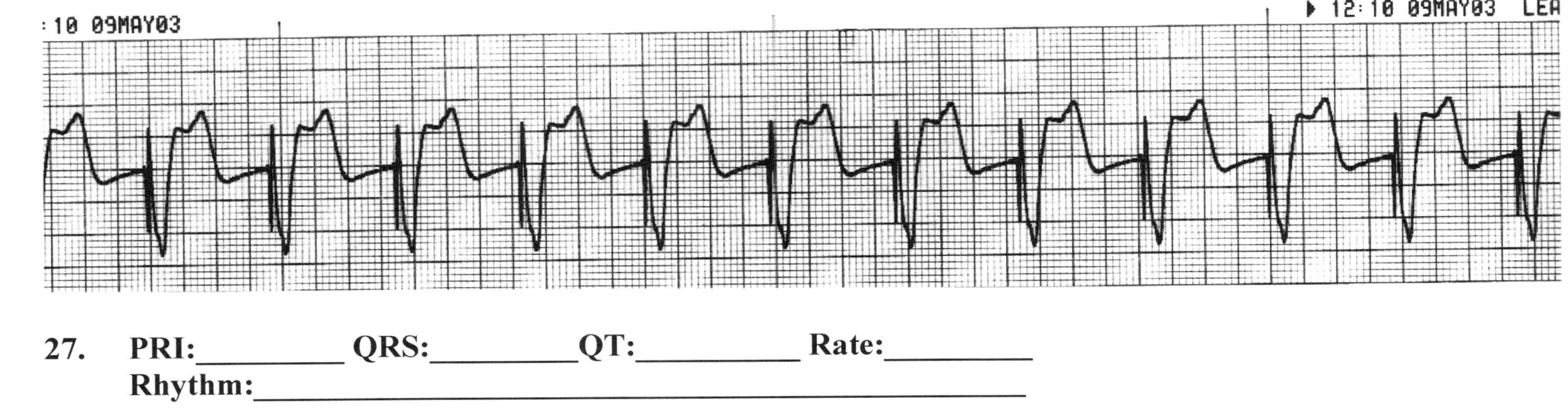

27. PRI:________ QRS:________ QT:_________ Rate:________
Rhythm:____

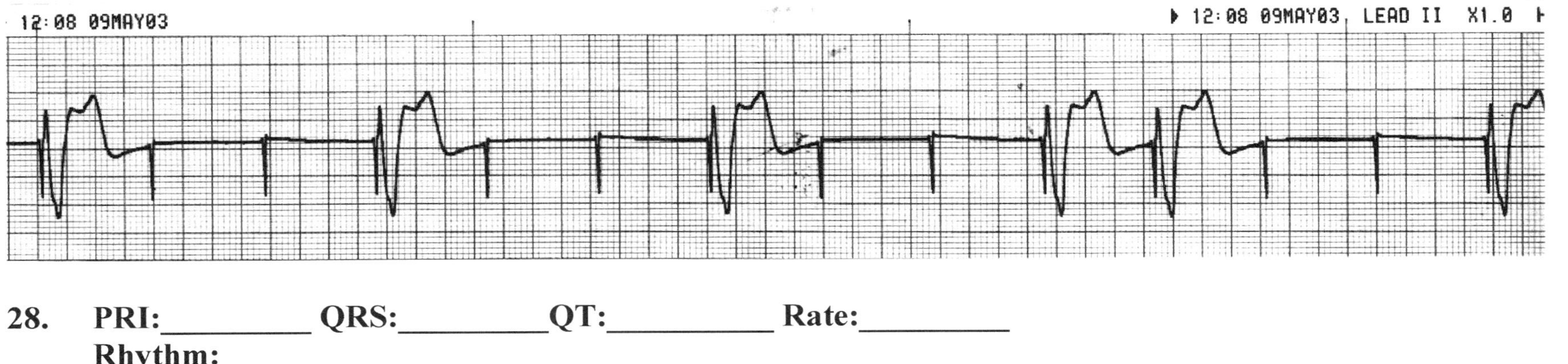

28. PRI:________ QRS:________ QT:_________ Rate:________
Rhythm:____

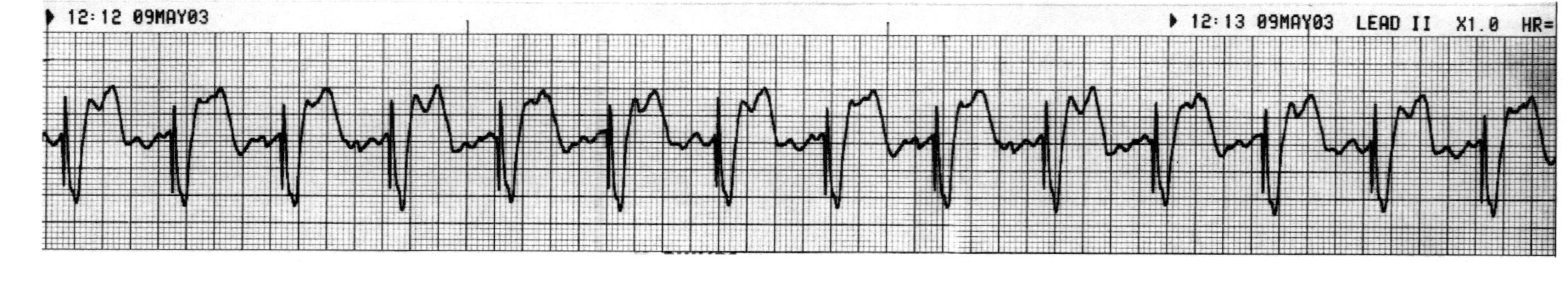

29. **PRI:_________ QRS:_________ QT:_________ Rate:_________**
Rhythm:____

Notes:

Putting it all together:

(These strips are tough, but if you are able to accurately interpret these strips, you are well on your way!)

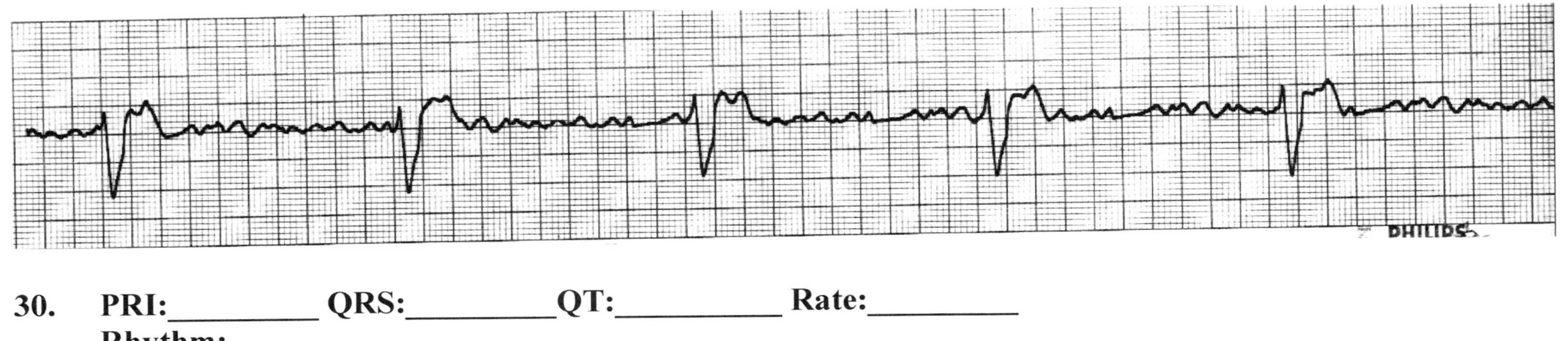

30. **PRI:**________ **QRS:**________ **QT:**_________ **Rate:**________

Rhythm:____

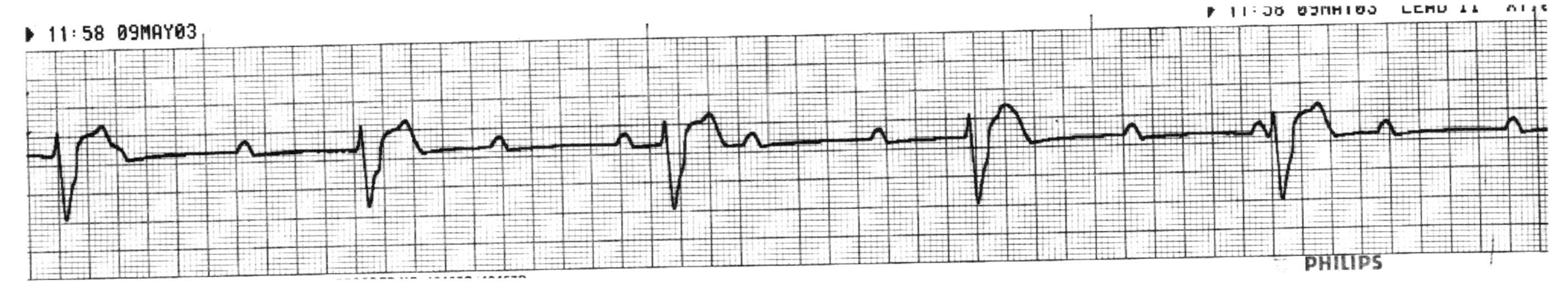

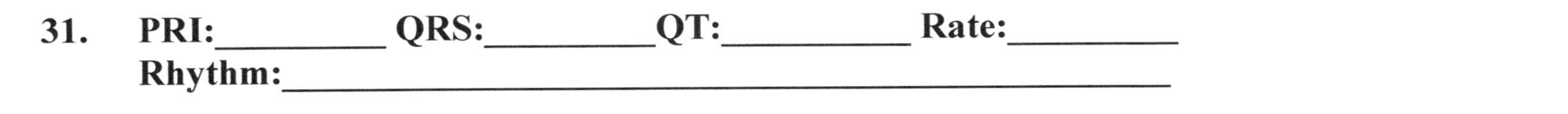

31. **PRI:**________ **QRS:**________ **QT:**_________ **Rate:**________

Rhythm:____

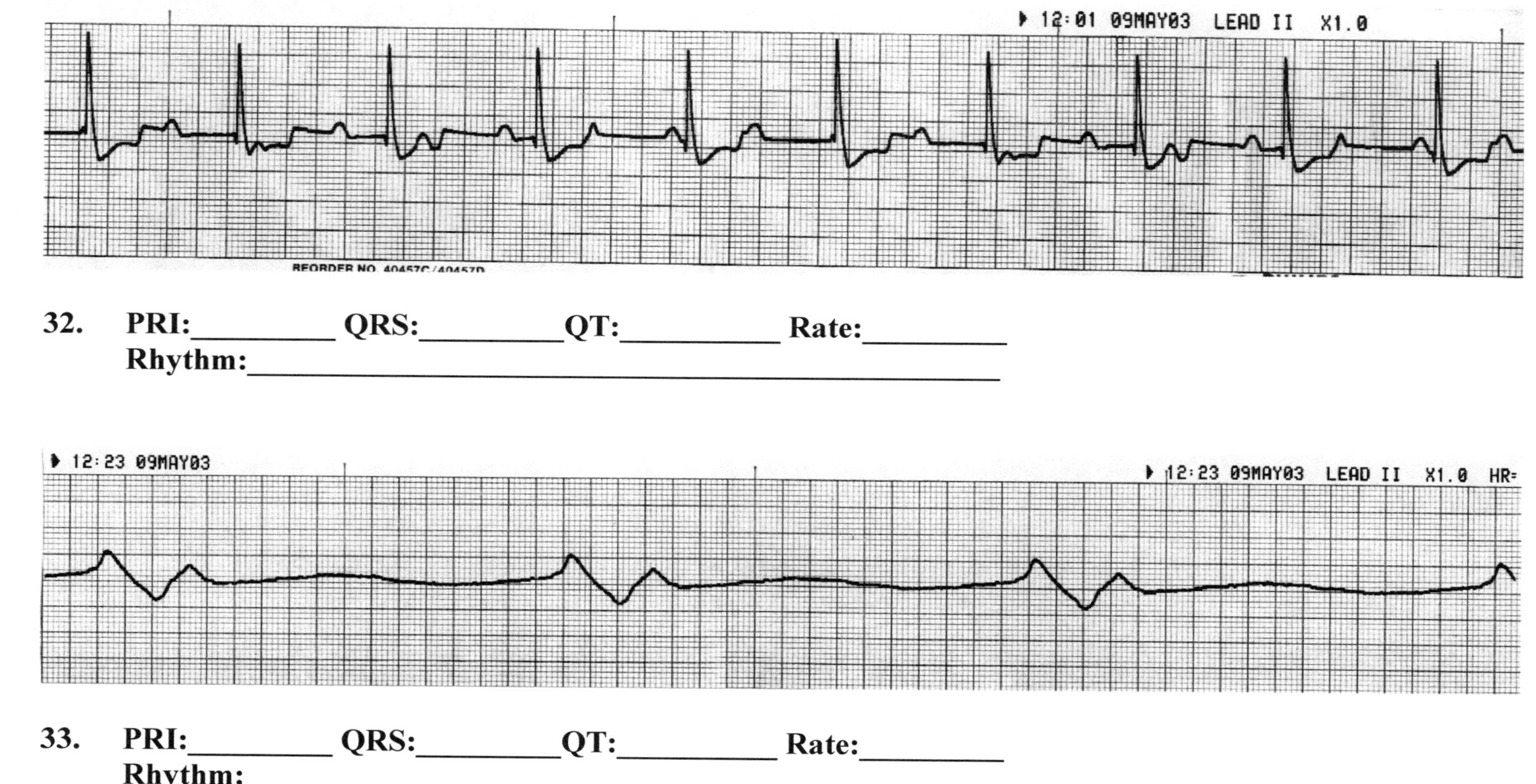

32. **PRI:_________ QRS:_________ QT:__________ Rate:_________**
Rhythm:____

33. **PRI:_________ QRS:_________ QT:__________ Rate:_________**
Rhythm:____

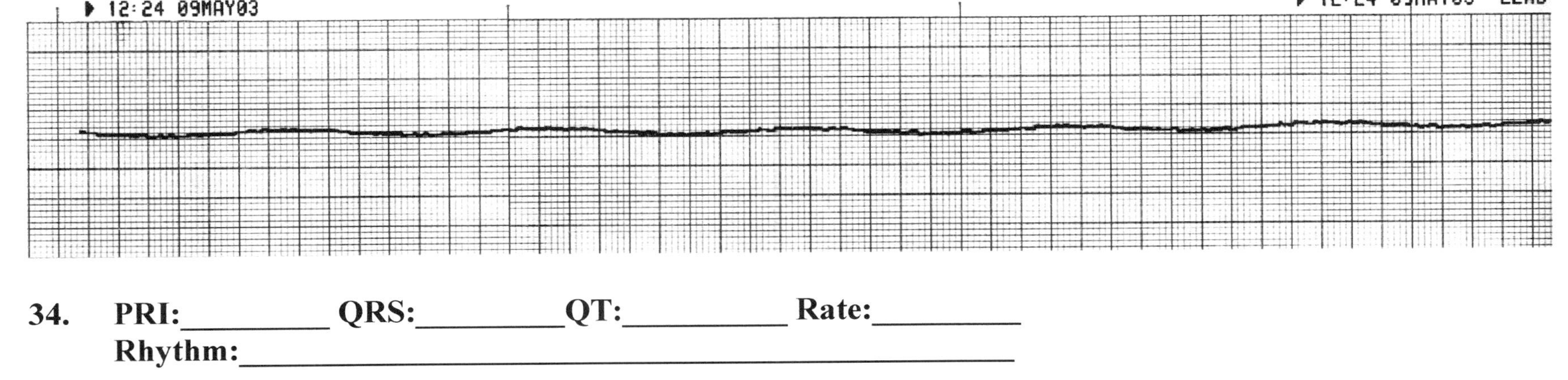

34. PRI:_________ QRS:_________ QT:__________ Rate:_________
Rhythm:____

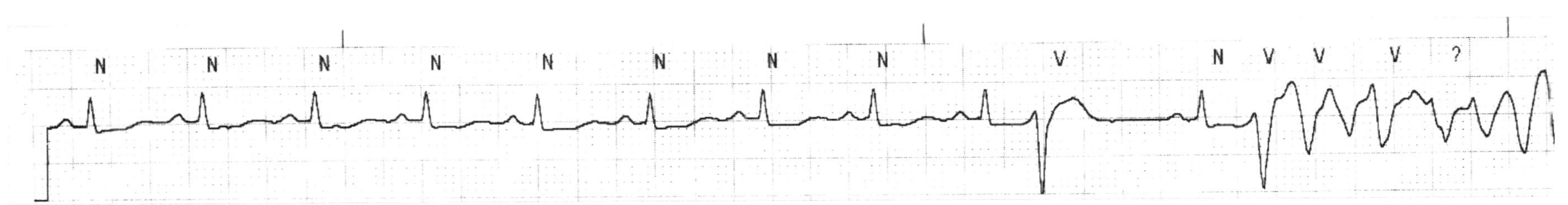

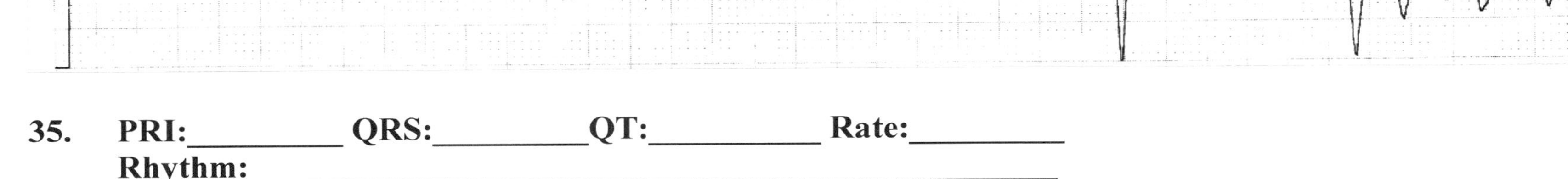

35. PRI:_________ QRS:_________ QT:__________ Rate:_________
Rhythm:____

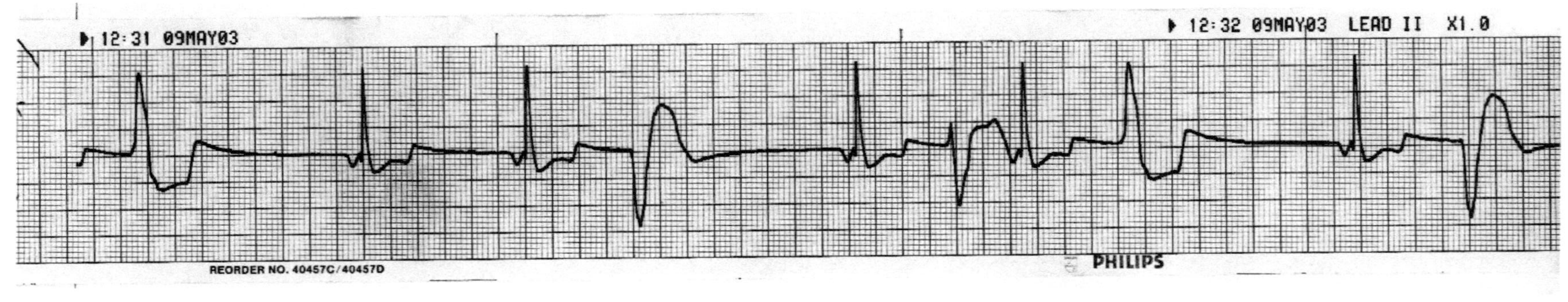

36. PRI:________ QRS:________ QT:_________ Rate:________
Rhythm:____

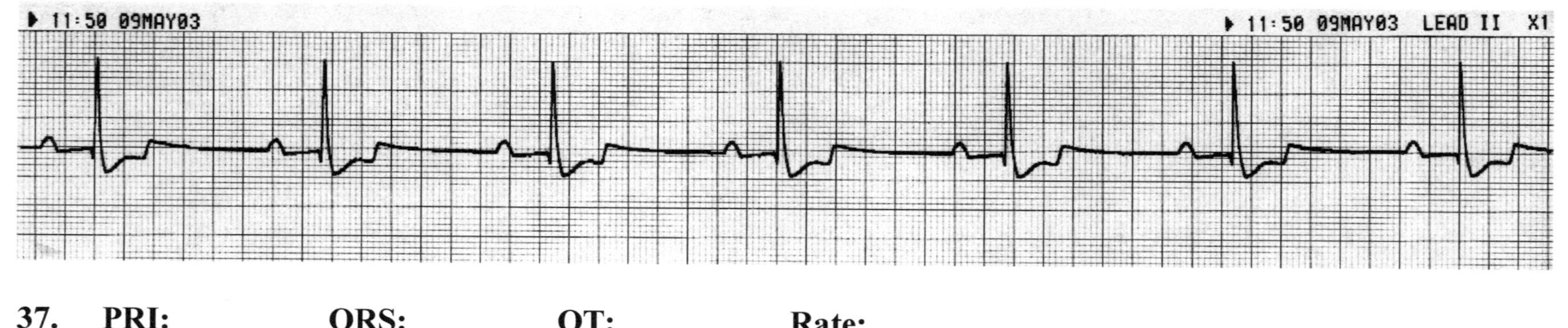

37. PRI:________ QRS:________ QT:_________ Rate:________
Rhythm:____

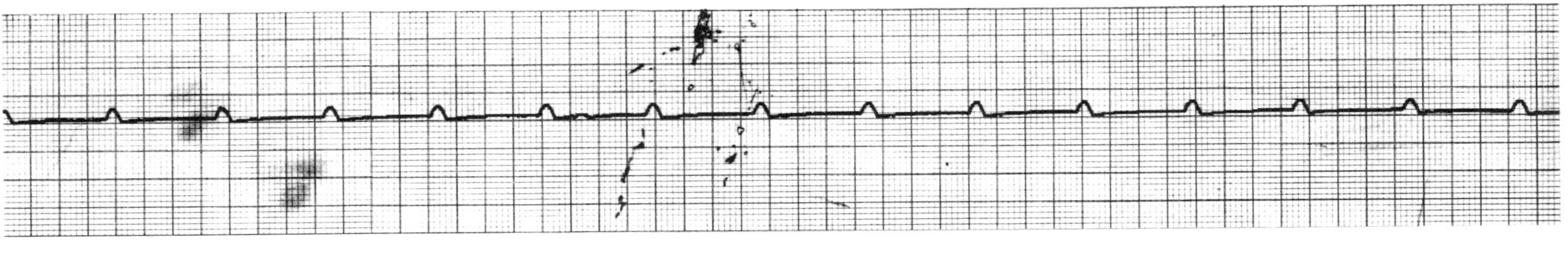

38. PRI:________ QRS:________QT:__________ Rate:_________
Rhythm:____

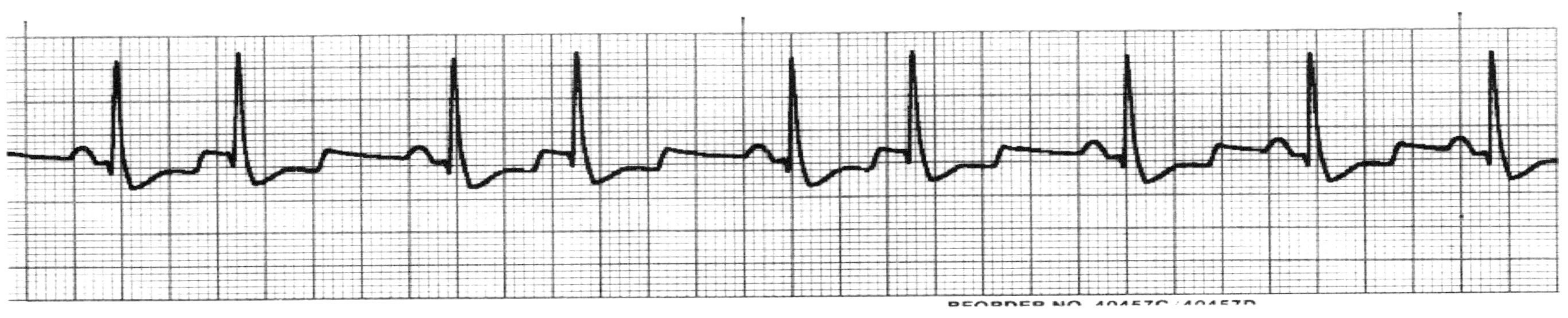

39. PRI:________ QRS:________QT:__________ Rate:_________
Rhythm:____

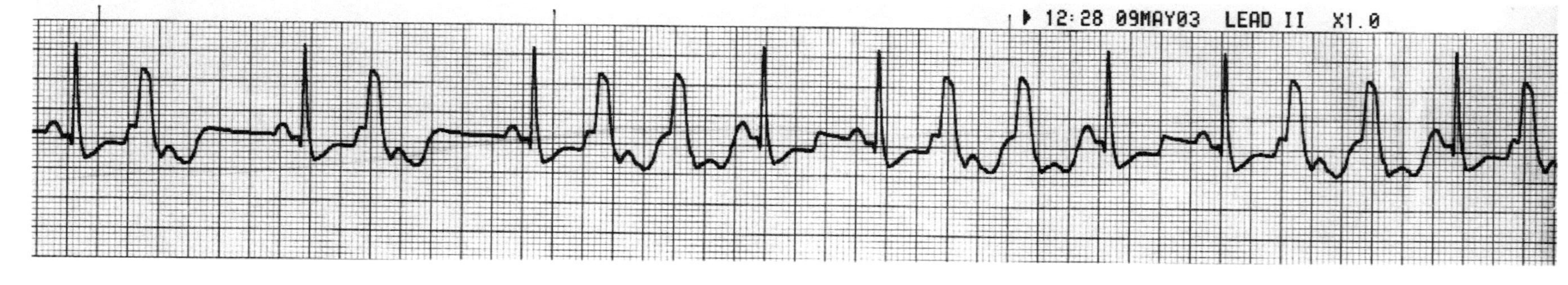

40. PRI:_________ QRS:_________ QT:__________ Rate:_________
Rhythm:____

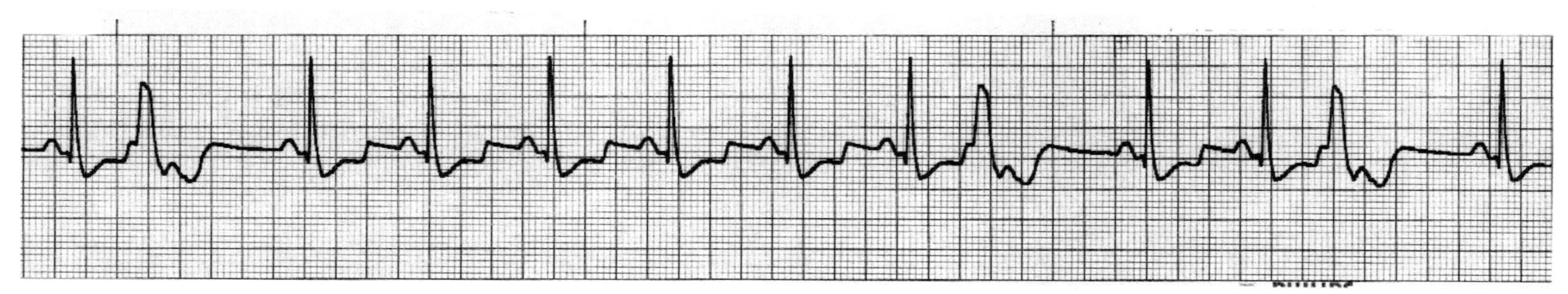

41. PRI:_________ QRS:_________ QT:__________ Rate:_________
Rhythm:____

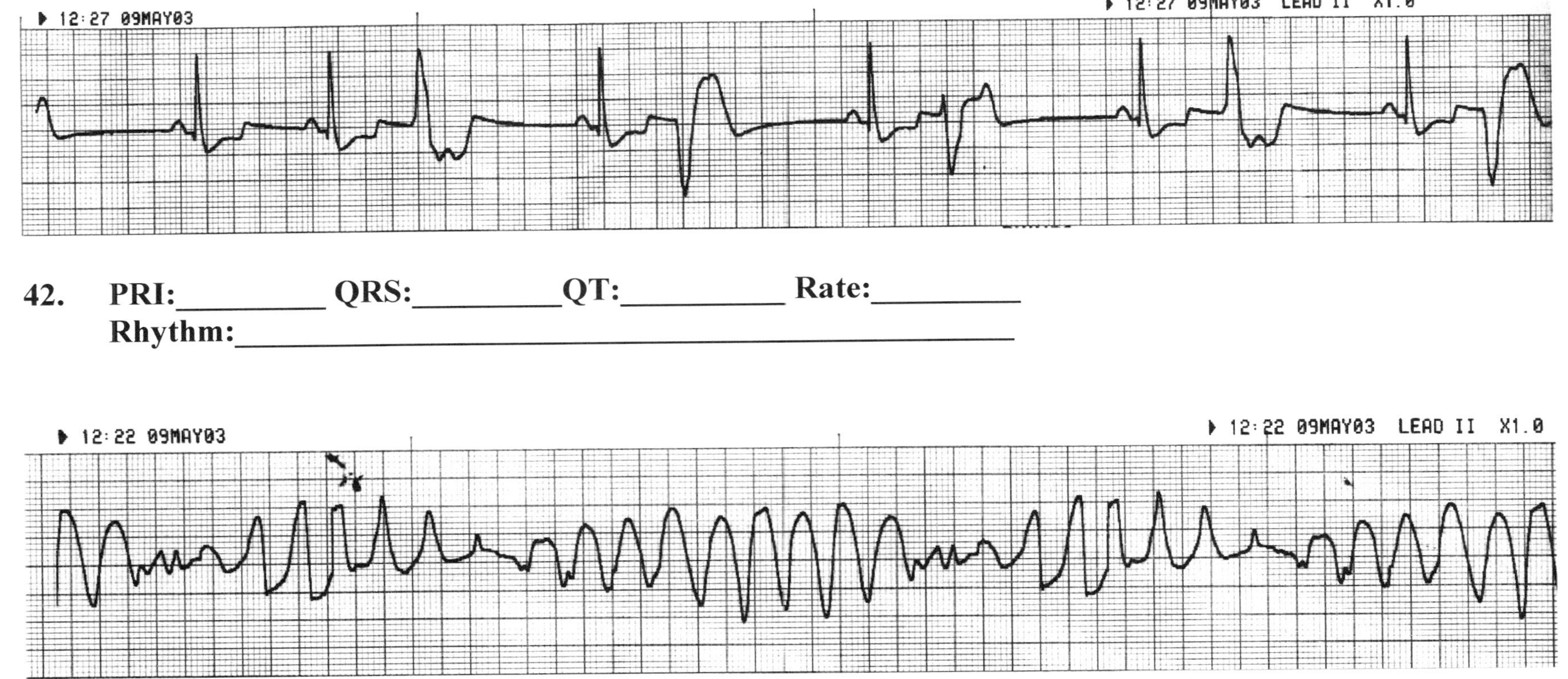

42. PRI:_________ QRS:_________ QT:__________ Rate:_________
Rhythm:____

43. PRI:_________ QRS:_________ QT:__________ Rate:_________
Rhythm:____

Notes

Workbook Answer Key

Cardiac Structures

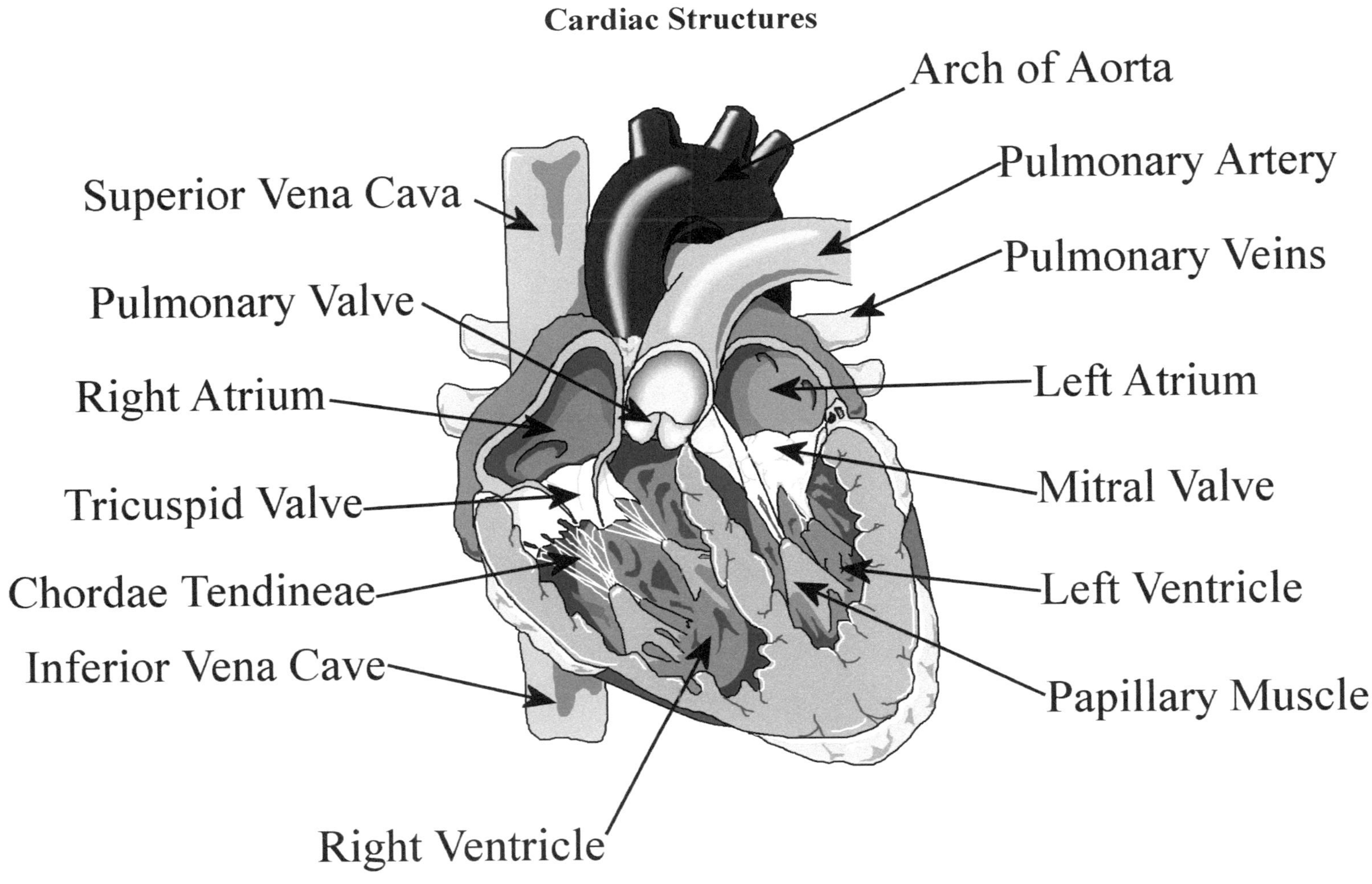

Cardiac Electrical System

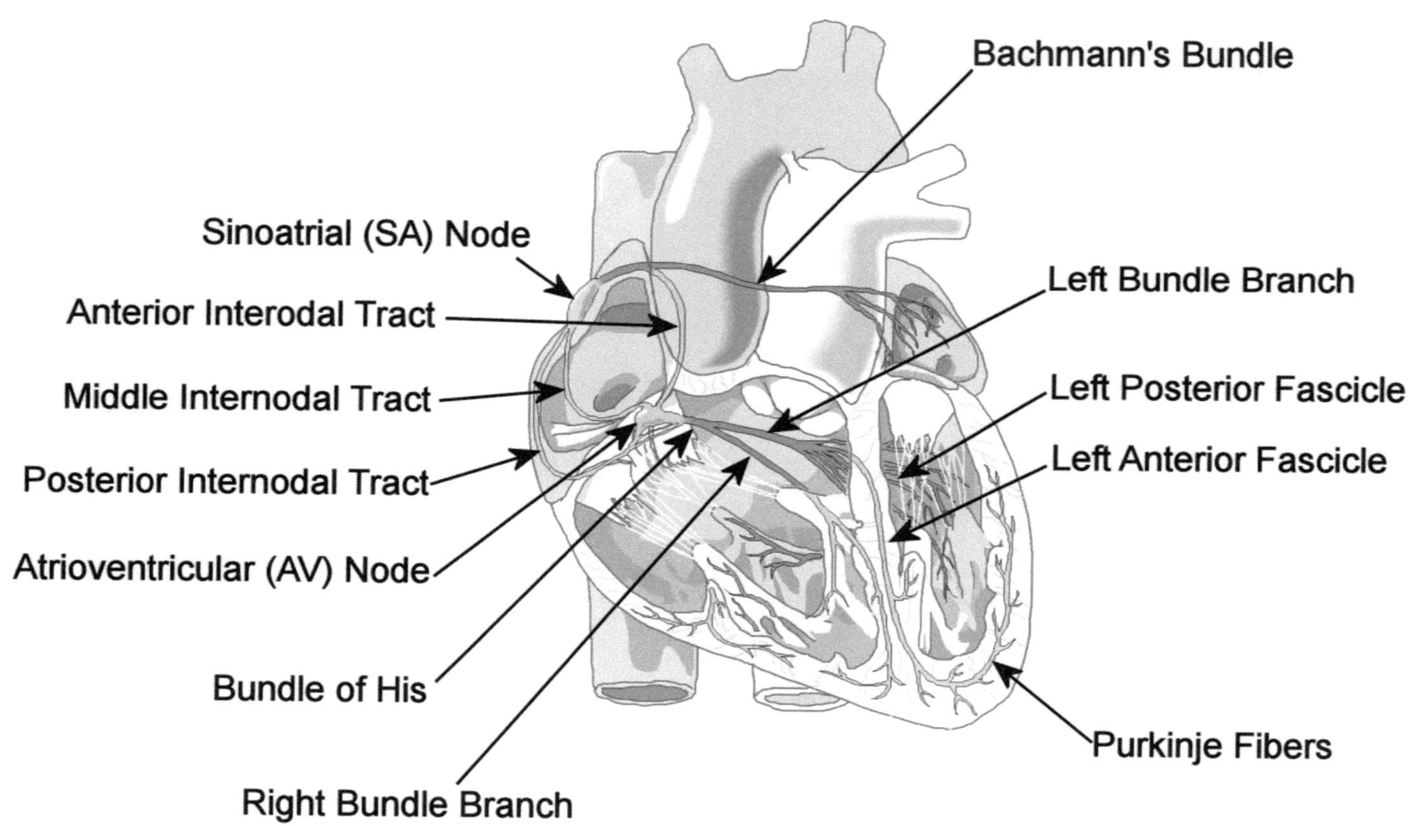

Coronary Arteries

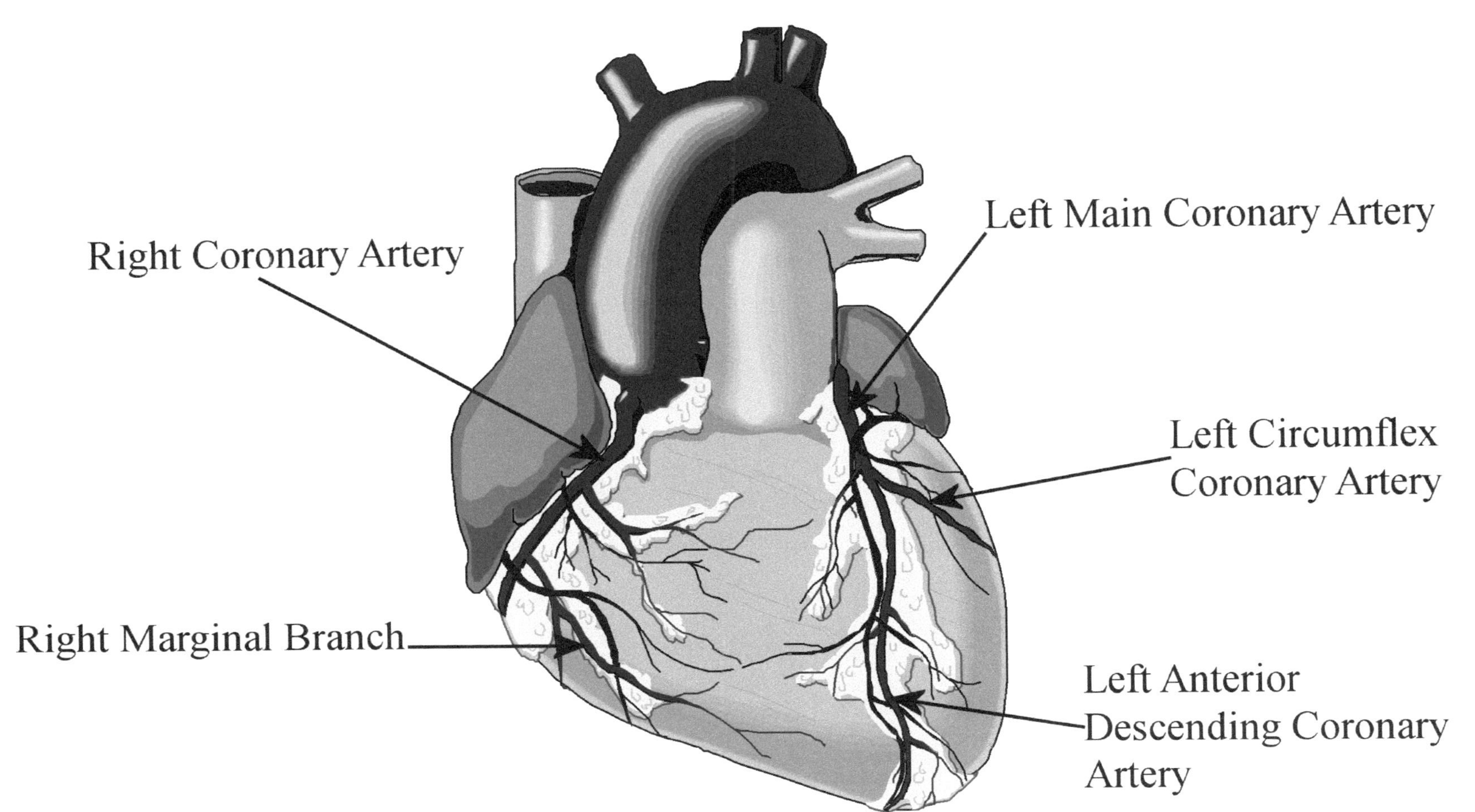

Basic EKG Workbook Rhythm Answer

1.
PRI: 0.18-0.20
QRS: 0.12
QT: 0.76
Rate: 50-75
Rhythm: Sinus Arrhythmia

2.
PRI: 0.16-0.18
QRS: 0.10
QT: 0.56
Rate: 80
Rhythm: Sinus with frequent Premature Atrial Contractions (PACs)

3.
PRI: 0.18
QRS: 0.10
QT: 0.84
Rate: 30-50
Rhythm: Sinus Brady/Arrhythmia

4.
PRI: 0.18
QRS: 0.10
QT: 0.80
Rate: 30
Rhythm: Sinus Bradycardia

5.
PRI: 0.18
QRS: 0.08
QT: 0.32
Rate: 120
Rhythm: Sinus Tachycardia

6.
PRI: 0
QRS: 0.10
QT: 0.60
Rate: 30
Rhythm: Junctional

7.
PRI: 0.14
QRS: 0.08
QT: 0.64
Rate: 50
Rhythm: Accelerated Junctional, P wave is inverted before QRS

8.
PRI: 0
QRS: 0.12
QT: 0.12
Rate: 78
Rhythm: Accelerated Junctional

9.
PRI: 0
QRS: 0.10
QT: 0.38
Rate: 120
Rhythm: Junctional Tachycardia

10.
PRI: 0
QRS: 0.08
QT: 0.32
Rate: 160
Rhythm: Atrial Fibrillation

11.
PRI: 0
QRS: 0.12
QT: 0.60
Rate: 25-100
Rhythm: Atrial Fibrillation

12.
PRI: 0
QRS: 0.08
QT: 0.32
Rate: 75-100
Rhythm: Atrial Fibrillation

13.
PRI: 0
QRS: 0.08
QT: 0.36
Rate: 75
Rhythm: Atrial Flutter 4:1

14.
PRI: 0
QRS: 0.18
QT: 0.60
Rate: 33
Rhythm: Idioventricular

15.
PRI: 0
QRS: 0.12
QT: 0.64
Rate: 54
Rhythm: Accelerated Idioventricular

16.
PRI: 0
QRS: 0.18
QT: 0
Rate: 290
Rhythm: Torsade de Pointes

17.
PRI: 0
QRS: 0.16
QT: 0.40
Rate: 120
Rhythm: Ventricular Tachycardia

18.
PRI: 0
QRS: 0
QT: 0
Rate: 0
Rhythm: Ventricular Fibrillation

19.
PRI: 0.16-0.32
QRS: 0.06
QT: 0.44
Rate: 42-75
Rhythm: 2nd Degree Atrial Ventricular Block Type I - Wenkebach

20.
PRI: 0.18-0.60
QRS: 0.08
QT: 60
Rate: 37-50
Rhythm: 2nd Degree Atrial Ventricular Block Type I - Wenkebach

21.
PRI: 0.16
QRS: 0.08
QT: 0.36
Rate: 60-85
Rhythm: 2nd Degree Atrial Ventricular Block Type II with frequent PAC's

22.
PRI: 0.18
QRS: 0.12
QT: 0.42
Rate: 40-75
Rhythm: 2nd Degree Atrial Ventricular Block Type II

23.
PRI: 0.16
QRS: 0.08
QT: 0.36
Rate: 50
Rhythm: 2nd Degree Atrial Ventricular Block Type II

24.
PRI: 0
QRS: 0.20
QT: 0.44
Rate: Atrial 75 / Ventricular 50
Rhythm: 3rd Degree Atrial Ventricular Block with Accelerated Idioventricular Rhythm

25.
PRI: 0.20
QRS: 0.08
QT: 0.40
Rate: 86
Rhythm: Atrial pacer with 100% capture

26.
PRI: 0.20
QRS: 0.20
QT: 0.44
Rate: 75
Rhythm: Dual chamber pacer with 100% capture

27.
PRI: 0
QRS: 0.18
QT: 0.44
Rate: 75
Rhythm: Ventricular Paced Rhythm with 100% capture

28.
PRI: 0
QRS: 0.16
QT: 0.44
Rate: 30
Rhythm: Ventricular paced rhythm with loss of capture

29.
PRI: 0
QRS: 0.20
QT: 0.48
Rate: 75
Rhythm: Ventricular pacer with 100% capture

30.
PRI: 0
QRS: 0.20
QT: 0.46
Rate: 30
Rhythm: 3rd degree heart block/A-Fib/Idioventricular Rhythm

31.
PRI: 0
QRS: 0.20
QT: 0.46
Rate: 30
Rhythm: 3rd Degree heart block with Idioventricular Rhythm

32.
PRI: 0
QRS: 0.08
QT: 0.44
Rate: 60
Rhythm: 3rd Degree heart block with Junctional Rhythm

33.
PRI: 0
QRS: 0
QT: 0
Rate: 18
Rhythm: Agonal rhythm of a dying heart

34.
PRI: 0
QRS: 0
QT: 0
Rate: 0
Rhythm: Asystole

35.
PRI: 0.16
QRS: 0.08
QT: 0.36
Rate: 95
Rhythm: Sinus with R on T PVC going into Ventricular Tachycardia

36.
PRI: 0.16
QRS: 0.08
QT: 0.52
Rate: 70
Rhythm: Junctional with frequent multi-focal PVCs

37.
PRI: 0.36
QRS: 0.08
QT: 0.56
Rate: 40
Rhythm: Sinus with 1st Degree Atrial Ventricular Block

38.
PRI: 0
QRS: 0
QT: 0
Rate: 0
Rhythm: Ventricular Standstill

39.
PRI: 0.16
QRS: 0.08
QT: 0.48
Rate: 80
Rhythm: Sinus with Premature Atrial Contractions

40.
PRI: 0.18
QRS: 0.08
QT: 0.36
Rate: 90
Rhythm: Sinus with frequent single and couplet unifocal PVCs

41.
PRI: 0.18
QRS: 0.08
QT: 0.44
Rate: 80
Rhythm: Sinus with frequent unifocal PVCs

42.
PRI: 0.20
QRS: 0.08
QT: 0.56
Rate: 60
Rhythm: Sinus with frequent multiform PVCs

43.
PRI: 0.20
QRS: 0.18
QT: 0
Rate: 200
Rhythm: Torsade de Pointes

EKG Pocket Guide

The pocket guide on the following pages can be cut out and folded for a quick and handy reference. Included in this EKG pocket guide are the following:

- Methods to calculate regular and irregular heart rates.
- EKG graph paper measurements.
- All major EKG rhythms and their measurements.
- You are granted permission to copy the EKG pocket guide for your personal use only.

EKG Rhythms At A Glance

Presented By
Mind-Forge Education

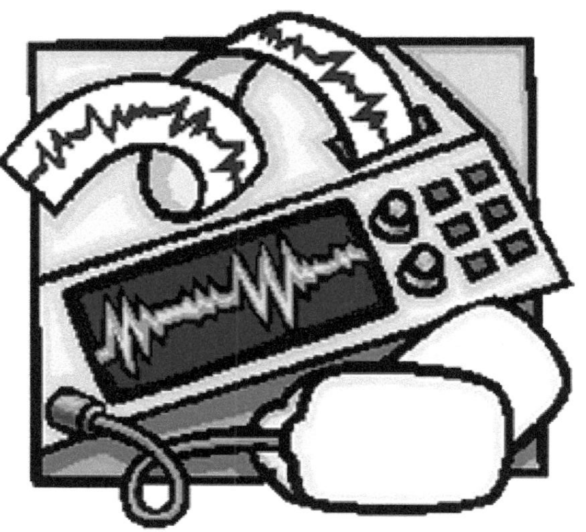

Created By
Douglas Martin MSN RN
Mind-Forge Education Services
www.mind-forge.com

Normal EKG And Measurements

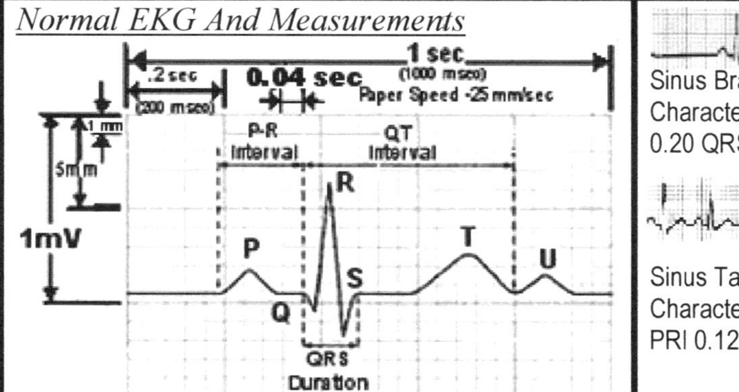

Calculation of Heart Rate: Regular Rhythm: Count the number of LARGE boxes between R waves and divide into 300.

Irregular rhythm: Count the number of R waves in a six second and multiply by 10.

EKG Strips

Normal Sinus Rhythm.

Characteristics: Regular Rhythm. Rate 60-100 PRI < 0.12-0.20
QRS < 0.12 QT < 0.40

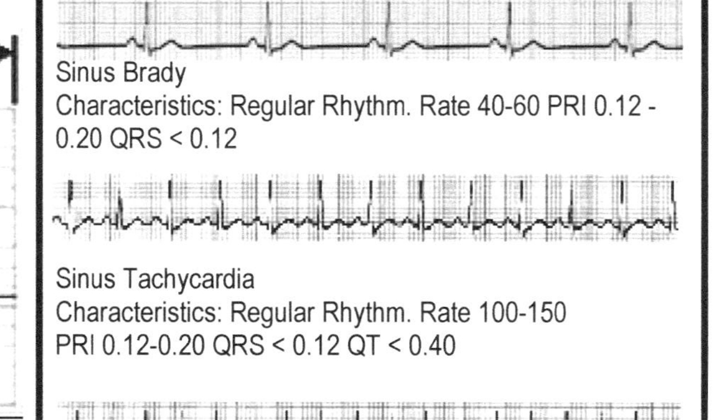

Sinus Brady

Characteristics: Regular Rhythm. Rate 40-60 PRI 0.12 - 0.20 QRS < 0.12

Sinus Tachycardia

Characteristics: Regular Rhythm. Rate 100-150
PRI 0.12-0.20 QRS < 0.12 QT < 0.40

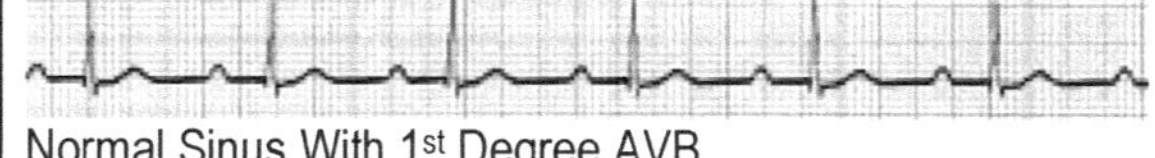

Atrial Tachycardia AKA SVT

Characteristics: Regular Rhythm. Rate 150-250
PRI 0.12-0.20 QRS < 0.12 QT < 0.40

* *If rate is very fast, P wave may not be visible.*

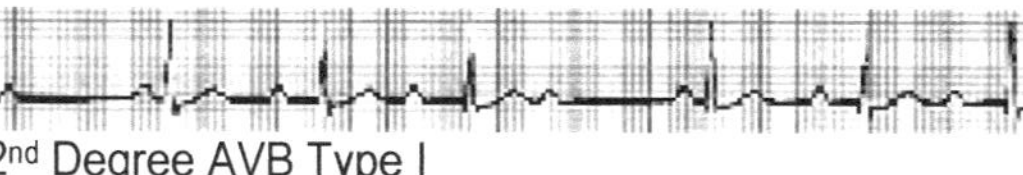

Normal Sinus With 1st Degree AVB

Characteristics: Regular Rhythm. Rate 60-100 PRI > 0.20
QRS < 0.12 QT < 0.40

2nd Degree AVB Type I

Characteristics: Irregular Rhythm. Rate 60-100 PRI *
QRS < 0.12 QT < 0.40

* PRI becomes progressively longer until a QRS is dropped. P wave present with no QRS conducted.

2nd Degree Heart Block Type II

Characteristics: Regular or Irregular*. Rate 60-100

PRI 0.12-0.20 QRS < 0.12 May be wider if BBB present.

*Rhythm may be irregular if a variable conduction block is present. With conducted beats, PRI can be normal. Every 2nd. 3rd or 4th beat is block at the AV node.

3rd Degree AVB or Complete Heart Block

Characteristics: Rhythm Regular or Irregular. Rate Depends on site of pacer. PRI Variable. QRS Variable. QT Variable.

* Rate can be 60-100. Usually underlying rhythm is Junctional escape at 40-60 or ventricular escape at 20-40. PRI is not constant. QRS width depends on pacer site. Junctional Escape QRS <. 012. Ventricular escape QRS is 0.12 or greater.

Atrial Fibrillation

Characteristics: Irregular Rhythm. No discernable P waves or PRI 0. QRS < 0.12 QT < 0.40

* Atrial activity may be coarse or fine at the base line. These undulations are called F waves.

Atrial Flutter

Characteristics: Rhythm/Rate Variable or Regular.

PRI 0. QRS < 0.12 Qt < 0.40

* P waves have a saw tooth appearance called F waves. More than one F wave is present for each QRS. Conduction can be 4:1, 3:1 or 2:1.

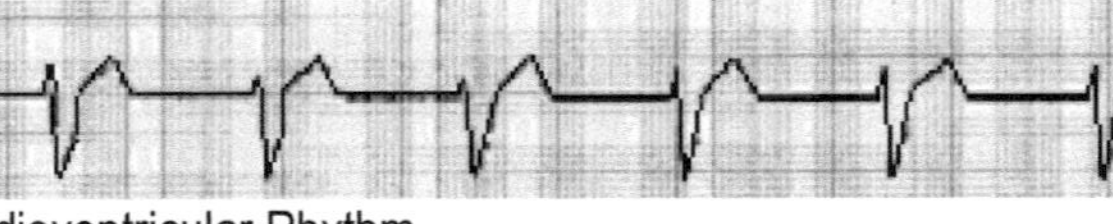

Junctional Rhythm with retrograde P wave conduction.

Characteristics: Regular rhythm. Rate 40-60.

PRI * Usually 0. QRS < 0.12 QT < 0.40

* P wave may be absent, inverted before the QRS, buried in the QRS or inverted behind the QRS.

Lethal Dysrhythmias

Ventricular Tachycardia. AKA VTACH

Characteristics: Rhythm Regular Rate 100-250 PRI: 0 QRS: 0.12 or greater. May be very wide or bizarre.

*** Is this stable or unstable?**

Ventricular Fibrillation AKA VFIB

Characteristics: Rhythm *Total Chaos.* Rate 0. PRI 0. QRS 0. No discernable PQRST complexes.

Torsades de Pointes *(Twisting of the points)*

Characteristics: Rhythm may be regular or irregular.

Rate 150-250. PRI 0 QRS Morphology Changes QT > 0.40

* Anything that lengthens the QT may cause Torsades.

Asystole

Characteristics: Straight line on monitor with no PQRTS complexes noted. ***Always confirm in another Lead.***

Idioventricular Rhythm

Characteristics: Rhythm Regular. Rate 40 or less.

PRI 0 QRS 0.12 or greater Qt < 0.40

* A ventricular rate of 41-100 is noted as "Accelerated Idioventricular Rhythm"

Paced Rhythm with 100% Capture

Characteristics: Pacer may be dual or single chambered, fixed rate or demand. For each spike, there should be a QRS response. Spikes without a QRS response denotes loss of capture.